ITALY'S MANY DIASPORAS

Donna R. Gabaccia

University of Washington Press
Seattle

First published 2000 in the UK
by UCL Press
11 New Fetter Lane, London EC4P 4EE
and in the United States of America
by University of Washington Press
PO Box 50096, Seattle, WA 98145-5096

UCL Press is an imprint of the Taylor & Francis Group

Typeset in Garamond by BC Typesetting, Bristol
Printed and bound in Great Britain by
St Edmundsbury Press, Bury St Edmunds, Suffolk

Library of Congress Cataloguing-in-Publication Data
Gabaccia, Donna R., 1949–
Italy's many diasporas/Donna R. Gabaccia.
p. cm. – (Global diasporas)
Includes bibliographical references and index.
ISBN 0–295–97917–8 (alk. paper). – ISBN 0–295–97918–6 (alk. paper)
1. Italians – Foreign countries. 2. Italy – Emigration and immigration. I. Title. II. Series.
DG453.G33 2000
909′.0451–dc21 99-42484
CIP

In memory of my father, Remo Palmo Gabaccia, 1926–97, and commemorating the centennial of the fatti di maggio, Milan, 1898

CONTENTS

TABLES

FOREWORD BY THE SERIES EDITOR

Over the last few years and even since this series was launched in 1997, the expression "diaspora" has diffused into general social scientific discourse. This has revived old meanings and developed new understandings of the migration experience in cultural studies, history, geography, sociology, and politics. But, just as in the catwalks of Milan and Paris, fashion is no certain guide to utility or to ultimate acceptance. As more and more scholars use the notion of "diaspora" we have rapidly to learn what are its strengths and limitations.

In this respect Donna Gabaccia is an informed and helpful skeptic. I have provided a fuller definition and discussion of diasporas in the introductory book to this series (R. Cohen, *Global Diasporas: An Introduction*, 1997) but, in brief, diasporas are partly about the scattering of people. Certainly, many millions of people left Italy — perhaps 2 million in the period 1790–1870 and over 26 million thereafter, when the newly founded Italian state started counting. However, calling them "Italians" was problematic as no single nationhood existed. This leads to the author's first rupture with the classic diasporic tradition. Regional, city, and village identities remained powerful among Italian migrants, so it seems sensible to talk, as Gabaccia does, of diasporas in the plural, rather than assume there is a single Italian people that dispersed to many places.

The migration from Italy was also generally "voluntary," though the force of adverse economic circumstances should always qualify that term. Unlike the classic "victim diasporas" (the Jews, Armenians, and Africans providing the paradigmatic cases), most migrants from Italy left in pursuit of trade or work. This is, however, no particular bar to understanding them as members of a diaspora or several diasporas, as the victim tradition was too narrow a formula. Diasporas are also about continuing connections with "home." And we can and should understand "home" metaphorically as well as literally. There is a trite saying that captures this point. "Home," it is said, "is where the heart is" and perhaps we should add "where the head is" and also (given the global adoption of Italian food) "where the taste buds are."

The continuation of circulatory migration, of Italian and sub-Italian identities, of the Italian language abroad as well as the development of "Little Italies" in several countries shows that the cultural and affective elements of a diasporic condition were highly developed among Italy's migrants. The sharing of Italianate Catholicism and the occupational niches colonized by many workers of Italian origins also enhanced solidarities. What Gabaccia calls "the rebellious poor" of Italy filled the global labor market with stonecutters, masons, weavers, spinners, waiters, machine operators, and many other occupations besides. The specificity of their employment, the experience of migration and remigration and their continued cultural distinctiveness meant that Italians constantly shuttled between their cosmopolitan, national, and local experiences. They were often prominent members of the labor movements in their countries of settlement and perhaps just as often organized among their own kind exclusively. Gabaccia's book provides a major contribution to understanding the dynamics of working-class organization among those of Italian origins.

One of the aspects of diasporic consciousness that is only recently being systematically considered is the attitudes and activities of the sending state with respect to their nationals abroad. Often seen as a "drain" or even as "betrayers" of their country, nationals abroad are frequently cast in a negative light. However, such were the numbers, political influence, and economic power of Italians abroad that the Italian state sought to capture their loyalty. The contest between radical and fascist traditions in the diaspora is brilliantly addressed in this book. Italy now receives as many immigrants as it loses in emigrants, while its national identity (precarious enough in the pre-1870 period) has also been fundamentally altered by its incorporation into the European Union. For these reasons the contours of state–diaspora relations have changed again, a configuration usefully discussed in the final two chapters.

In this book we have a prominent scholar at the height of her powers writing a scholarly, yet accessible and theoretically innovative book. I am happy to commend it to the reader and delighted we have secured it for this series.

Robin Cohen
Series Editor

PREFACE

Ironically, this book began with my decision to write no more about Italians. By 1988, I had completed two books on Sicilian migration to the U.S., but I had also run out of scholarly steam. "I always wanted to write books," I had scribbled in frustration one day while riding the train home to Westchester from the New York Public Library, "but not books that no one reads." The American audience for books on Sicilian migrations seemed frustratingly small. With its relentless focus on the incorporation of immigrants into the mainstream of American life, U.S. historiography found room only for stories of the making of Americans and of American ethnic groups. I was instead interested in human mobility and in connections among countries and cultures.

But there was one final Italian-American writing obligation facing me, and I was looking forward to finishing it in 1990. That final task was to prepare a paper on Italian migrants in labor movements around the world for the meeting of the American Italian Historical Association in New Orleans. When I gave this, my "swan song" paper, I also met new colleagues — Fraser Ottanelli, Joseph'Barton, Sam Baily, Carol Bonomo Albright, and Fernando DeVoto — who praised the global and comparative approach I attempted in it. Collectively, their enthusiasm worked its influence on me. Rather than write another book about immigrants for American historians, I began to fantasize of writing a book about Italian migration for scholars interested in class and ethnicity in many lands. I was about to discover that these formed quite a sizeable group. I also quickly met an interdisciplinary group of Americanists searching for global and comparative ways to interpret modern history and life in the U.S. Here was the audience I wanted to address.

Eight years later, I have chosen to address it by writing a general overview of Italy's migrations, in which theoretical exegesis takes second place to descriptive and empirical material. I have used endnotes to introduce readers to the literature and to broader theoretical issues without cluttering the text with either. Only in this fashion, I am convinced, can the inherent drama of migratory life in the modern world emerge to capture a reader's imagination.

Within six months of the New Orleans conference, Fraser Ottanelli and I began to plan what I fondly called the "Italians Everywhere" project. In early 1993, we circulated the first of six newsletters on "Italian Workers Around the World." Our mailing list eventually grew to 150 people in fifteen countries. The project culminated in an April 1996 conference in Tampa, Florida.

This book is one of several that grew out of collaboration among correspondents of the "Italians Everywhere" project. The interpretations offered in this book (and its errors) are mine alone. Still, I could (and would) not have written it without the support and contributions of a very large group of people whom I am pleased to acknowledge and to thank again here.

My first thanks, of course, go to Fraser Ottanelli, with whom I organized many scholarly panels, wrote an article, organized a conference, and edited a collection of essays. That collection focused on the lives of migratory Italian radicals who firmly believed, as we did too, "For Us There Are No Frontiers." Fraser's deep knowledge of and heartfelt commitment to anti-fascism as an international movement were essential to the success of the project. I also offer special thanks to Franca Iacovetta, with whom I also wrote an article and with whom I am currently editing a collection of essays on Italian women in the diaspora. In this book I draw on what I learned from both my co-authors but I have tried to craft a narrative distinct from the ones I created with either.

In the early stages of the "Italians Everywhere" project, some newsletter correspondents were particularly helpful in extending our circle. They contributed to this book by sending published materials, alerting me to bibliographical aids, and sharing the names and addresses of interested colleagues. For early help, I especially thank Dora Barrancos, Jonathan Barzman, Antonio Bechelloni, Lucia Birnbaum, Paola Corti, Isabella Felici, Franca Iacovetta, the late George Pozzetta and Gianfausto Rosoli, Carina Silberstein, Matteo San Filippo, Chiara Vangelista, Rudy Vecoli, and Eric Vial. Since many of them are themselves migrants, or descendants of migrants, it is worth noting that they live in Italy, Argentina, Canada, France, and the U.S.

I would like also to thank the many people who read and

responded critically to an early, and very long, review essay in which I first tried out some of the ideas expressed in this book. Readers have included Sam Baily, Antonio Bechelloni, John Davis, Alex DeGrand, Nancy Green, Nando Fasce, Pietro Rinaldo Fanesi, Emilio Franzina, Fraser Ottanelli, Franco Ramella, and Elisabetta Vezzosi. Additional feedback came from audiences at the Giovanni Agnelli Foundation and Fondazione Einaudi in Turin, at the Calandra Institute and Queens College, at New York's City University, and at the Center for Migration Studies in Buenos Aires. I appreciate the help of Maddalena Tirabassi, Phil Cannistraro, and Fernando DeVoto, who arranged my visits to these three important centers for the study of Italian migrations.

I am also grateful to the many people outside Italian studies who pressed me to examine the Italian case in comparative perspective, and to make it comprehensible to non-Italianists. Several pointed to the concept of "diaspora" as a fresh, if problematic, way to think about international migratory networks. Robin Cohen, editor of the Global Diasporas series, first put the thought in my mind, and helped me see the social scientific possibilities of the concept. Khachig Tölölyan (editor of *Diaspora*) helped me recognize scholars using the concept in anthropology, literature, and cultural studies. A number of meetings with David Thelen (editor of the *Journal of American History* and initiator of most of its projects for "internationalizing" U.S. history) pushed me further, to consider how far global methodologies could challenge national historiography. A conference on "States and Diasporas" organized by Mexicanist and sociologist Robert Smith of Barnard College raised new questions by comparing the Italian case to those of present-day migrants from the Philippines, Latin America, and the Caribbean. My best friends in comparative history — notably Dirk Hoerder, Sam Baily, and Nancy Green — continued to ask me difficult methodological and conceptual questions. I am not sure I have responded to all their skeptical queries but I gave them my best try.

I thank also the many people who read parts of the manuscript. Some of them were close to home and included my colleagues John Smail, Lisa Lindsey, and Jane Laurent, along with participants in the History Department Seminar in Charlotte. (Here in Charlotte, Lyman Johnson also helped me fill sizeable gaps in my knowledge of the twentieth-century history of Argentina.) Farther afield, Sam Baily, Peter D'Agostino, Enrico Del Lago, Dora Dumont, Nancy Green, Dirk Hoerder, Franca Iacovetta, Jennifer Guglielmo, Walter

Nugent, and Linda Reeder gave me some very useful suggestions on earlier drafts.

Financial support in the form of research and travel grants came from the Charles H. Stone Fund and from the Faculty Grants Program, both at the University of North Carolina at Charlotte.

I sadly dedicate this book to my father — my remaining personal connection to my own family's small corner of one of Italy's many proletarian diasporas. I regret that he did not live to see the book completed but I appreciate his trust that I would finish it, even while we both knew he was dying. I hope I have inscribed his personal example — as hard-working "testa dura," as restless man and migrant, and as an italiano nel mondo — on every page.

As always, my largest thanks go to those precious friends who sustained me during my years of restlessness and globetrotting. Where would my life be without Dorothy, Jeanne, Jeffrey, Tamino, and Thomas? Everywhere, but nowhere.

INTRODUCTION

Residents of Italy have been leaving home from "time immemorial."[1] But what do we know of the consequences of that mobility for the migrants themselves, for Italy, or for the many countries worldwide where they settled? We do not even know their total numbers. From 1870 on, over twenty-six million people officially declared their intention to leave Italy. We know the figure because a new Italian state was troubled enough about the possible consequences of emigration to bother to count them. For most of the previous 800 years, Italy had been little more than a "geographical expression" — a peninsula at the center of first the classical world, and then of the Renaissance of a new western civilization. Active participants in the cultural and economic transformations that created modern European civilization, the residents of Italy lacked any common government, language, or culture. There was no Italian nation or Italian people before 1861. An infinitesimally small group of nationalists first imagined a national community of Italians and then created an Italian national state in 1861.[2] Italy was still such an abstraction to Sicilian peasants that when they heard "il Mille" (the 1000 supporters of Garibaldi, who invaded the island) crying "Viva Garibaldi; viva l'Italia!" they assumed "Talia" was Garibaldi's wife. Italy's national state quickly took on the task of nation-building. But even today, some scholars firmly maintain there is no such thing as "an Italian."[3] Without Italians, or an Italian nation, it is difficult to write with much confidence about an Italian diaspora, with its assumptions about connections among peoples living outside their homeland.

Yet every diaspora begins as a scattering of a people; it originates in human migration.[4] The residents of the Italian peninsula and its nearest large islands have been among the most migratory of peoples on the earth. A look at the map explains why: rarely did they live far

Table I.1 Origins of migrants from Italy, 1200–1975 (%)

	*1200–1500**	*1500–1790**	*1790–1875**	*1876–1915*†	*1916–45*†	*1946–75*†
North	44	42	47	45	49	28
Center	52	43	31	20	19	19
South‡	3	16	22	35	32	52
(Number	248	1195	681	14,037,531	4,482,347	7,335,081)

Sources:
* Elite migrants only; Ugo E. Imperatori, *Dizionario di italiani all'estero (dal secolo XIII sino ad oggi)* (Genoa: L'Emigrante, 1956).
† Gianfasuto Rosoli (ed.), *Un secolo di emigrazione italiana, 1876–1976* (Rome: Centro Studi Emigrazione, 1976).

Note
‡ Includes Sicily and Sardinia.

from the sea, and most of their lands were mountainous, places where sufficient food, room, and shelter were attained only with considerable effort. Noteworthy already in the late Middle Ages, migrations from Italy assumed epic proportions during the modern era. Roughly two million people migrated between 1790 and 1870. Fourteen million more applied to their new government to leave in the years between 1876 and 1914. Another four million declared the same intention between 1916 and 1945. More than seven million left Italy between 1945 and 1975 (a number that excludes the nine million who traversed the very long cultural and geographic distance still separating Italy's deep South from its continental North during the same years).[5] Italy's earliest migrations were more culturally and economically than quantitatively significant. But subsequent migrations from Italy have been among the most important of the modern world. Italians made up roughly 10 percent of international migrations in the century between Waterloo and World War I.[6] The numbers leaving Italy after 1876 surpassed the population of the newly unified country in 1861.

Nor was migration a regional phenomenon characteristic only of isolated corners of the Italian peninsula. Initially greatest from the North and center of the peninsula — as Table I.1 suggests — long-distance travel became a defining characteristic of life everywhere in Italy after 1500, long before either a nation of Italians or a national state of Italy existed. Over the course of the nineteenth century, the most numerous migrants were those from the Veneto (in the northeast) and from Campania and Sicily (in the South). Only slightly smaller numbers left Piedmont, Lombardy, and Friuli (in the northeast and northwest) and Calabria (in the South). It was these regions, and their many villages, towns, and cities — not an Italian nation — that produced most of Italy's many diasporas.[7]

The modern diasporas of Italy were webs of social connections and channels of communication between the wider world and a particular paese (village) or patria (hometown). They rested on migrants' close identification with the face-to-face communities of family, neighborhood, and native town. This combination of loyalties left but little space for a nationalism that put identity with a nation or with a national state above all others. It is no accident that the modern Italian word for country is the same as its word for village (paese). Nor is it accidental that the modern Italian words for citizenship (cittadinanza) and citizen (cittadino) originally defined loyalty to a city, not to a nation (nazione or nazionalità), people (popolo), or race (stirpe or razza). Patriotism (love of the patria or birthplace)

Table I.2 Destinations of migrants from Italy, 1200–1975 (%)

	*1200–1500**	*1500–1790**	*1790–1875**	*1876–1915*†	*1916–45*†	*1946–75*†
Asia	24	9	4	–	–	–
Africa	6	3	6	–	3	1
Europe	67	84	77	44	52	68
North America	1	1	4	31	25	13
South America	1	3	7	24	19	13
Australia	–	–	1	–	1	5
(Number	248	1195	681	14,037,531	4,482,347	7,335,081)

Sources:
* Elite migrants only; Imperatori, *Dizionario di italiani all'estero.*
† Rosoli, *Un secolo di emigrazione.*

continues to give Italian loyalties today a localism that distinguishes them from other modern forms of nationalism.[8]

Rooted to particular home places, migrants from the Italian peninsula have nevertheless scattered rather widely around the world (see Table I.2). Although other parts of Europe were always the most important destination for those leaving Italy, migrants from Italy have consistently traveled to more than one continent. First, in 1200–1500, they migrated around the Mediterranean Sea (that linked Africa, Asia, and Europe) and across the Alps. In the years between 1500 and 1900, the residents of Italy moved both across the Mediterranean and the Alps and out into the Atlantic world; after 1900, they ventured into the Pacific world as well. Over 100 years, the vast territories of the United States received the largest group of migrants from Italy (over five million). It was followed closely by the much smaller countries of France and Switzerland (over four million each), Argentina (three million), Germany (2.5 million), and Brazil (1.5 million). North and South Africa and Australia also received smaller, but still significant, migrations from Italy.

With centuries of global scattering behind them, it is unsurprising that the residents of modern Italy often cite the proverb "tutto il mondo è paese" (all the world is one home place). Simultaneously, the proverb expresses the essential unity of humans and the smallness of the world, and it strongly suggests that a person can be at home anywhere. Scholars have long been aware of the cosmopolitan as well as local dimensions of Italian life. They have written of "the Italians in the world" (gli italiani nel mondo) and of the international connections created among "gli italiani fuori d'Italia" (Italians outside Italy).[9] Some scholars have instead analyzed an "Italia fuori d'Italia" (Italy outside of Italy), pointing to the existence of a global and cosmopolitan culture produced in Italy — a civiltà italiana.[10] In a very different but equally global view of modern life, Marxist scholars have tended instead to view modern Italian migrants as members of an international proletariat.[11] In this book, I focus repeatedly on the connections between the cosmopolitan and international ("il mondo") and the local and personal ("il paese") in defining Italy's unique history of migration.

My goal in this book is to assess the significance of the global networks formed by Italy's many migrants for the history of Italy and of the countries where migrants worked and settled. These networks certainly resemble diasporas, and a few scholars have called them that.[12] I will argue that migration rarely created a national or united

Italian diaspora. But it did create many temporary, and changing, diasporas of peoples with identities and loyalties poorly summed up by the national term, Italian. There have been diasporas of people from Sambuca (in Sicily) and the Biellese area (in Piedmont, in Italy's northwest). There have been diasporas of merchants, chimneysweeps, and hurdy-gurdy men from specific Italian towns. And there have been diasporas of Italian-speaking anarchists and Mussolini-inspired fascists. Only in the last few decades, however, can one speak of migrants from Italy united by a national identity, seeking and sustaining firm connections to a national government at home, or with a shared sense of identity around the world. To label most of Italy's migratory networks as diasporas forces us to accept a somewhat minimalist definition of a diaspora, and to detach it from its associations with nationalism.

Unlike the diasporas best known to Americans — those of Africans and Jews, groups which Robin Cohen terms "victim diasporas" — relatively few of the migrants who left Italy were victims. Most were not exiles. True, a minority fled to avoid political persecution, especially in the politically fractious years of the 1300s, during the movement to create an independent Italian state in the early nineteenth century, and again during the fascist era.[13] But most leaving Italy did so to improve their economic lot, not in response to political, religious, or cultural persecution. All migrants from Italy — even the political exiles — exercised considerable choice over when to leave home and where to go. Theirs was no involuntary or sudden scattering into an exile without end. Italy's diasporas more resembled that of the ancient sea-going and entrepreneurial Greeks (merchant diasporas) than that of enslaved Africans or persecuted Jews.

Psychologically this mattered enormously. Oppressed peoples create diasporic identities that originate with the traumatic experience of being forced from a beloved homeland to which they cannot easily return. Key to diasporic identities is a sense of loss; loss creates strong feelings of connection among those sharing it. Shared loss can also create communal commitments to reproducing that identity and passing it along to future biological generations — both to accommodate them to life in exile and to help them dream of ending it. In the modern era, the trauma and loss associated with involuntary scattering often generate a fervent desire to return again to the homeland, and to create there a territory of safety, often in the form of a national state. Of these characteristics of diasporic identities, migrants from Italy shared only a fervent desire to return to a beloved home. For migrants from Italy, home was a place — the patria or paese — not a people, nation, or descent group.

Evidence from the late Middle Ages down to the 1960s suggests that those who left Italy did so fully intending to return. But unlike the persecuted peoples of Africa or Jerusalem, the vast majority of migrants from Italy could do so with relative ease. Large numbers did in fact return to Italy. Before 1790, between 55 percent (in the politically chaotic 1600s) and 79 percent (in the more peaceful and prosperous 1400s) of Italy's most prominent migrants eventually returned.[14] Approximately three-quarters of the political exiles of the years between 1790 and 1860 also went home again.[15] For the modern period rates of return are slightly lower, about 50 percent.[16] Thus, the demographic consequences of migration from Italy have always been considerably less than the total volumes of migration suggest.[17] But this does not mean we should ignore the impact of generations of comings and goings on life in Italy. The large number of transients, not the smaller groups that settled abroad or returned home, defined the particularity of Italy's diasporas.

Male predominance in migrations from Italy is the most telling evidence of intentions to return. While declining over time, male majorities seem to have been a continuous pattern in migrations from Italy (see Table I.3). By contrast, in present-day migrations of refugees, as in the scattering of Jews and Armenians, women and men fled in equal numbers, or women predominated (because men had been killed in the ethnic conflicts that preceded the scattering). As a student of contemporary refugee movements notes, among those who flee, as in a shipwreck, the understanding often is "women and children first."[18]

Table I.3 Females among migrants from Italy, 1200–1975

	% female	*Total number of migrants*
1200–1500*	0	248
1500–1790*	2	1195
1790–1875*	3	681
1876–1915†	19	14,037,531
1916–45†	33	4,482,347
1946–75†	29	7,335,081

Sources:
* Elite migrants only; Imperatori, *Dizionario di italiani all'estero.*
† Rosoli, *Un secolo di emigrazione italiana.*

As long as they could return freely to their hometown, migrant "men without women"[19] had few reasons to create a national diasporic identity. Most migrant men left Italy with familial, local, or regional loyalties. These ties effectively linked migrants together into migratory networks or "chains" of mutual assistance.[20] If a sense of diasporic identity emerged among migrants abroad it did so first among the Biellesi or the Sicilians, and only more slowly among a group that came to think of itself as a nation of "italiani nel mondo."[21] Even when migrant men came to see themselves as Italians while living abroad, the women to whom they returned, and with whom many would then raise children, often did not. During periods of intensive male emigration, biological and cultural reproduction took place more often in Italy than abroad. When women left Italy, furthermore, it reflected a decision to settle more permanently abroad; women's migration facilitated reproduction in a new homeland, and signaled the beginnings of permanent incorporation in new nations.[22]

Evidence that the residents of Italy valued or felt themselves part of a nation of Italians at home or abroad is spotty even for the modern period. Identification with an Italian nation remained limited to urban, educated, and bourgeois persons — no more than 10 percent of the population — from the time of the Enlightenment until the very recent past. In the nineteenth century, this identification seemed to consist mainly of pride in the accomplishment of medieval merchants, Renaissance artists and humanists, and Catholic clerics. To feel Italian was to identify culturally with "civiltà italiana" — an elite culture that had developed in and spread from Italy to Europe between 1000 and 1600. Educated Italians took considerable pride in the civilizing mission of Italian culture. But civiltà italiana was a slim foundation on which to build a modern nation. It had little meaning for the mass of Italians, even in the late nineteenth century when a government investigator in the Marches on Italy's eastern coast commented dryly, "sentiments of 'italianità' ['italianism'] among our peasants are not very lively."[23] These same peasants and workers made up as many as 90 percent of the modern, mass migrations.[24] Most migrants did not speak Italian. They knew little of civiltà italiana. Many despised the Italian national state. These migrants were unlikely unifiers of Italy's many diasporas into a single national one.

Even among middle-class urbanites proud to identify with civiltà italiana and with the nation of Italians that produced it, contempt for the national Italian state — once it finally existed — was rampant.

Whether under constitutional monarchy, fascist dictatorship, or the postwar Italian republics, residents of Italy perceived their state, and its governments, at best, as little capable. At worst they saw them as thieves — "governo, ladro," a Sicilian proverb noted — who took taxes and draft-age sons while giving nothing in return.

Distrust of the state runs deep in Italian life, and it persists to this day — as anyone who studies contemporary Italian politics and life quickly discovers.[25] A source of cultural unity, distrust of the state scarcely encourages the development of a proudly shared national identity. Yet distrust of states proved as culturally durable among ordinary Italians as enthusiasm for civiltà italiana had been among Italy's bourgeois rulers: it deeply influenced migrants' behavior as both workers and citizens abroad. Irish, Greek, and Polish migrants often understood their migrations in search of work as a kind of exile; they enthusiastically supported nationalist movements at home.[26] Only rarely was this the case among migrants from Italy.

Although hedging the terms "Italian" and "diaspora" with many qualifications, I nevertheless believe it heuristically helpful to imagine the possibility of a single Italian diaspora. Most of what we know about Italy's diasporas has emerged in fragments[27] and recent efforts at synthesis have been superficial and uninformed.[28] We know much about immigrants from Italy in the U.S., Brazil, or a few other receiving countries. We know much about emigration from Biella or Sambuca, or from Friuli or the Abruzzi. We know almost nothing about connections among these places, or between them and their migrants abroad. National histories obscure the most outstanding characteristics of migrations from Italy — their worldwide dimensions and their circulatory character — and ignore the connections between Italy's villages, towns, and government and the "other Italies" (most Italian speakers called them "colonies") around the world.[29] The seductiveness of the term "diaspora" is that it forces us to look simultaneously at the many places to which migrants traveled, and at the connections among them.

Looking for an Italian diaspora — even if we do not find one — brings the global and circulatory character of migrations from Italy into sharp focus. It also forces us to ask what difference Italy's connections to the wider world have made to the national history of Italy and to the national histories of receiving countries. Would either Italy or the countries where Italians settled be the same had this migration not occurred? Were the experiences and identities of all Italians alike, regardless of where they lived? Clearly, the answer is "no" in both cases.

By asking how and in what ways migrations from Italy created diasporas we can see formation of class and national identities as processes common to national states around the world. We can also identify and compare variations in the emergence, growth, definitions, and hegemony of national loyalties in multi-ethnic populations. Finally, we can understand in a more precise and simultaneously more humane way the impact of international migrations on modern life. Scholars hope to represent worldwide migration in mind-numbing lists of statistics and multi-colored arrows thrusting across maps of the world. The study of a diaspora focuses attention instead on migrations' impact on human culture and identity, and on the evolution of the human collectivities — nations, states, families, neighborhoods, and home communities — that make life both human and culturally diverse.

By examining migrations globally we can see, for example, that the concepts of civiltà italiana and of an Italian nation emerged within, and were shaped by, the diasporas of their respective eras. We see as well how migratory workers in the modern period in some respects resembled an international proletariat, and we can see the impact of their comings and goings on national labor movements around the world. Not surprisingly, the limits, and the successes, of labor internationalism also take on new meaning when examined within the global networks of migrants.

In addition, we can see that both sending and receiving countries were without exception culturally diverse — one might choose to call them multi-ethnic — populations with relatively new national states. As such, they all faced a common challenge — to transform transient workers, along with racial minorities, indigenous peoples, and peasants, into citizens they could trust to fight for national interests as states defined them. The Italian nationalist Massimo D'Azeglio may never have uttered the sentiment so often attributed to him in the aftermath of Italy's unification — "Italy is made, now we must make Italians." Still, the task of creating loyal citizens out of transient workers was a huge one both in Italy and in the other countries where they labored. There were many nation-building states in the era of international migrations. The history of modern migrations from Italy is in large part a history of state efforts to incorporate migrants into multi-ethnic nations. A nation of emigrants like Italy thus shares much with the "nations of immigrants." A global perspective reveals why so many national states came to see migration as a problem threatening their stability in the early years of the twentieth century and how modern, multi-ethnic nations then developed sharply contrasting models of citizenship. Stephen Castles and Mark Miller have labeled these models racial or "ethnic" (as in Germany), unitary or "republican" (as in France), and "multi-cultural" or plural (as in the U.S.).[30]

Finally, viewing Italian migrations as a potential diaspora reveals the deep historical roots of what anthropologists today call "transnationalism."[31] Transnationalism is a way of life that connects family, work, and consciousness in more than one national territory. Migration made transnationalism a normal dimension of life for many, perhaps even most, working-class families in Italy in the nineteenth and twentieth centuries. Family discipline, economic security, reproduction, inheritance, romance, and dreams transcended national boundaries and bridged continents. Nor was transnationalism limited to the migrants. Italy's governing classes — both before and during fascism — sometimes dreamed of building an empire of migrants and the "colonies" they created abroad. Even after fascism's fall, Italy's government continued to tinker with bilateral treaties, definitions of citizenship, and forms of political representation that could build Italy's national strength through its migrants.

In short, transnationalism is no invention of a late twentieth-century or postmodern world. Nor is it the new threat to nation-state hegemony that cosmopolitan scholars (among whom I include myself) might hope. In the history of Italy's migrants, transnationalism appears as a recurring dimension of life in every world system, and an "ethno-scape" of ancient origins.[32] We can see early forms in the Mediterranean world first described by Fernand Braudel and in the Atlantic economies of the years 1500–1914, as well as in today's European Union.[33] Nation states emerged as powerful organizers of human life during an important era of transnationalism. Whether as sending or receiving countries, nation states have prospered while living with complications posed by their citizens' transnationalism. They have even actively encouraged transnationalism as a source of national power. In all likelihood, they can continue to do so.

While their lives were transnational, the "italiani nel mondo" did not form a "nation unbound," or a "de-territorialized nation state," as some scholars describe contemporary migrants, especially from the Caribbean.[34] National identities are arguably firmer and more central to international migrants today than they were to migrants from Italy in the early twentieth century. Even after 100 years of nation-building in Italy, cultural distrust of the state, combined with a love of the home village or city, remains important among the residents of Italy. With only a few exceptions, these sentiments prevented Italy's many migrants from thinking themselves, or behaving like, a nation unbound.

The chapters below examine four distinctive moments in the long history of migrations from Italy. Chapter 1 focuses on Italy's earliest long-distance migrants, between the late medieval era and the French Revolution, when merchants, students, clerics, and artists created the first "nations" from Italy and gave precise meaning to the adjective "Italian." Chapter 2 examines the diaspora of elite supporters of the Risorgimento (the Italian national movement) and their efforts to incorporate humble labor migrants into that nation, through diaspora nationalism. Chapters 3–6 examine the village-based proletarian diasporas[35] formed by millions of laborers migrating in and out of Italy between 1870 and 1940. Chapter 3 describes the economic globalization of the nineteenth century, and the place of workers from Italy — both male and female; both sedentary and migratory — within it. Chapter 4 describes the social connections of migrants and those in Italy, and the working-class (and transnational) way of life they created. Chapters 5 and 6 examine two distinctive phases in the nationalization of migrant workers. Between 1870 and World War I labor movements competed with Italian nationalists and the Catholic church to build "other Italies" that would support their very diverse causes. Chapter 6 shows how national states in turn pressured migratory workers to settle and to develop firmer national loyalties between the two World Wars, fomenting an ideological battle between fascists and anti-fascists over the values and character of the Italian nation. Chapters 7 and 8 examine the postwar world, the transformation of Italy from a sending to a receiving country, and the very diverse identities of migrants' descendants in multi-ethnic nations on four continents.

Whether at home or abroad, the legacy of centuries of migration from Italy is still visible today. Migration continuously reinforced the formation of complex, plural identities among Italians and their descendants. Some states successfully discouraged that pluralism. But in Italy, and in many receiving countries too, the national component of citizens' identities is far from being the only important one. For peoples still personally connected to the last two waves of migration from Italy, national loyalties often coexist in a creative tension with cosmopolitan and international experience. Equally creative but complex is the tension between national loyalties and attachments to family, local community, and region. Those tensions — not the art of the Renaissance — are at the heart of the civiltà italiana of Italy's modern and proletarian diasporas.

These tensions surely explain why a national Italian diaspora developed only as migration ceased. Ironically, however, we cannot grasp the significance of connections among the global and the local, the local and the national, or the national and the international until we actually take the time to consider the italiani nel mondo. Even a diaspora that never was can tell us much about the making of the modern world and of a modern Italy.

1

BEFORE ITALIANS: MAKING ITALIAN CULTURE AT HOME AND ABROAD

E tanti sun li Zenoexi
e per lo modo si'destexi
che und'eli van e stan
un'altra Zenoa ge fan.
(So many are the Genoese, scattered worldwide, that they build other Genoas wherever they reside.)[1]

In European languages, a nation (nazione in modern Italian) is a people sharing a sense of common origins, history, language, or culture. To the vernacular poet of the Middle Ages (above), the Genoese were a people (a natio, he might have called them).[2] Genoa had its own language that resembled Catalan (of northern Spain) as much as modern Italian. It had its own government, too — a republic. Genoa was a city state, a political form typical of the European late Middle Ages and much smaller than the larger nation (or national) states that developed later. Thus, most English speakers today would call the Genoese an ethnic or regional group. But in other languages, and in Balkan Europe today, groups no larger or culturally more distinct than the Genoese assert claims as nations to their own states.

In the days of our vernacular poet, Italy was neither a nation nor a state, but a "geographic expansion" — a peninsula that divided the Mediterranean into two halves. There were no Italians living there.[3] The noun "Italian" (originally from the Latin *italicus*) had been a label outsiders applied to many peoples of the Roman Empire centuries before. (These same peoples, even under Rome, called themselves something else — Etruschi, Sanniti, Liguri, Veneti, Galli, etc.) The term "Italian" largely disappeared from Italy as the Roman Empire shattered. By the year 1200, dozens of city states divided Italy's North and much of its center. Near the center of the peninsula, Rome was the seat of a universal church with close, if conflictual, ties to a Holy Roman Empire ruled politically from France or Germany. It still claimed the loyalties of all Christians although it had already lost the orthodox Christians of the Balkans, Greece, and the eastern Mediterranean. Along its southern coasts and islands were the western territories of the Eastern Roman (or Byzantine) Empire, ruled from Constantinople. Elsewhere in the South and nearby islands, centuries of Arab, Catalan, and Norman invasions produced ever-changing puzzles of interlocking colonial pieces. While the Arabs who dominated much of the Mediterranean saw the inhabitants of Italy along with other Europeans as Franks or Latins, human identities in 1200 in Italy were instead local.[4] Italians never called their group of immediate kinsmen a natio, and they most commonly identified with their hometown, calling it their patria (literally, fatherland) or paese (village), not their natio. A noteworthy departure from an imperial past, Italy's political fragmentation in 1200 was nevertheless typical of all of western Europe at the time.

But 550 years later, Italy had diverged sharply from its northern and western neighbors. Its inhabitants had not evolved into a single nation with a national state. The inhabitants of Spain, Portugal, and France had arguably formed nations already in 1500; the formation of a united nation of Britain, incorporating Ireland and Scotland, occurred between 1690 and 1707. By contrast, Italy, along with Germany, remained fragmented, and they did not create nation states until the second half of the nineteenth century.[5]

Still, both the identities and the political boundaries of Italy had changed in the intervening centuries. By the time of the French Revolution, four duchies (Milan, Tuscany, Parma, Modena) and four nominally republican city states (Genoa, Venice, Lucca, San Marino) dominated much of northern and central Italy. Three larger dynastic states (Savoy — later the Kingdom of Sardinia — of the northwest, the Papal States of the center, and the Kingdom of Naples and Sicily in the South) divided the rest of the peninsula and linked it to nearby islands. City states had given way to larger territorial units, to regional dialects, and to "nations" of Sicilians, Tuscans, and Marchegiani (from the Marches along the Adriatic Sea).

While migration did not cause either Italy's political fragmentation or its gradual evolution from city states to regional dynasties, the mobility of its residents certainly influenced the long and complex process of nation-building on the peninsula. When residents of Italy left home for long periods, or traveled long distances — as did the merchant mariners of Genoa — they often lived with former neighbors and called themselves a natio for the first time. When artists and architects left Italy to build and to decorate the courts and churches of Europe, they instead gave new meaning to the term Italian ("of Italy"). Long-distance migrations produced the first of Italy's diasporas — of missionaries and merchants, artists and musicians, and of the nations of Genoese, Lombards, and Florentines — without producing a united Italian nation or a politically united Italy.

Surprisingly, the residents of Italy exercised their greatest cultural influence on the wider world while they were most divided. Educated Europeans by 1500 recognized the existence of a distinctive civiltà italiana (civilization or culture originating in Italy).[6] They admired Italian art, music, science, architecture, humanist scholarship, and urban pleasures. Civiltà italiana changed Europe's place in the world, contributing to an economic and cultural renaissance of a once moribund and backward west.[7]

Historians have long extolled Italy's formative influence on European civilization, with Jacques Le Goff referring to the "waves of italianism" rolling periodically out of the peninsula. Fernand Braudel instead saw Italy radiating "its splendid thousand-colored light well beyond its own confines."[8] While striking, these images are misleading: ideas rarely travel independently of people. Civiltà italiana effectively exercised its civilizing influence precisely because so many migrants carried it from their hometowns to the courts, marketplaces, and universities across the mountains and seas that surround Italy. Their diasporas cast long shadows over all subsequent migrations from Italy and over all Italian interpretations of them.

From Italian primacy to "the Indies of the court of Vienna"

In the year 1000, western Europe was a backwater ("the land of the Franks on the western seas," as the Arabs saw it) on the periphery of a commercial economy centered in the Mediterranean.[9] There, the Arab caliphates of North Africa competed with the Byzantine Empire to control access (via "oriental" overland routes and the Indian Ocean) to the older, richer, and more politically advanced civilizations of South and East Asia.[10] After 1000, Italy became the geographic hinge on which economic and cultural innovation shifted gradually from east to west.[11] From 1200 to about 1500, Italy's city states played central roles in creating a new, dynamic economy around the Mediterranean. Its commercial cities of Venice (on the Adriatic) and Genoa (on the Tyrrhenian Sea) battled to control shipping and trade, and to acquire colonies or influence in both halves of the Mediterranean. Cities like Milan and Florence became important centers for banking, industry, and trading with transalpine Europe. These were the years of Italy's primacy, both culturally and economically.[12]

Historians debate when to date the onset of Italy's decline into economic stagnation and political dependency,[13] but the years between 1450 and 1550 seem a reasonable choice. Beginning in the 1400s, newly important dynasties in north and western Europe began to send explorers and traders (not a few of them from Italy) out into the Atlantic. They created both global empires for themselves and a new, and Europe-centered, world economy.[14] Unlike those of Amsterdam and the city states of the North Sea Hanseatic League, the merchants of Venice did not venture far out into the Atlantic. Nor did the residents of Italy coalesce into a larger nation around a single dynasty. Unlike Spain, Portugal, France, or England, none of Italy's many peoples built a sizeable mercantilist empire aimed at extracting and transferring the wealth of far-flung colonies. While other European nations conquered native civilizations in the Americas and enslaved Africans to work there, Italy became a minor dependency in Spain's and France's global empires. By 1700, Austria's European and dynastic, but nevertheless imperial, policies shaped life in much of Italy. Italy's transformation was astonishing: a late medieval metropole, it became an early modern dependency, the Austrian counterpart of Dutch and British colonies in the Indies. This transformation shaped the making of civiltà italiana, the migrations that spread it through the world, and the formation of nations in Italy itself.

Civiltà italiana first emerged during the period of economic expansion and relative political independence in Italy that historians of economy and politics label Europe's first commercial revolution. Italy was at the heart, and its residents became the key mediators, of an international economy centered on the Mediterranean but still drawing on the wealth and learning of the eastern Mediterranean Levant (or "Orient"), India, and China. Engaged in the slave trade, the spice trade, and the cloth trade, the merchants of Venice purchased Asian and local products from their Arab and Byzantine neighbors in the eastern Mediterranean and North Africa. They and their fellows then transported these goods, along with goods produced in Italy, to western and transalpine Europe, just emerging from its own Dark Ages.[15] Undoubtedly geography helps explain Italy's important mediating role. As a peninsula, it stretched almost to Africa, facing both the eastern and western shores of the Mediterranean; it was also the Mediterranean terminus to the most important Alpine passages from the north and west of the European sub-continent.

Italy's growing economic power emerged from its cities, largely — though not exclusively — in its northern and central sections.[16] Sea power made mariners and then merchants of the Genoese, Venetians, Pisans, and (in the South) Amalfitani; other cities like Messina, Milan, and Florence successfully transformed small-scale production into flourishing export industries in silk, metallurgy, arms, and woolen cloth. Genoa, Florence, Milan, and Venice specialized in banking. Rome, while widely recognized as the center of universal Catholicism, and of an alternative, central, and western but religious, culture, remained apart from most of these economic changes. The Papacy was in turmoil in the late Middle Ages — with popes reigning from France for part of the period. Before engaging in the missionary expansion that accompanied the building of European empires, several generations of Catholic clerics first had to struggle against the Germanic rulers of the Holy Roman Empire to speak unchallenged as heads of a united Christendom in western Europe.

Economic and urban growth in Italy went hand in hand. Cities expanded largely because of continuous in-migration from the countryside. However economically attractive, Italy's cities were unhealthy places; death rates remained high and urban populations seemed unable to reproduce themselves. Peasants and rural craftsmen plotted their moves to cities flourishing behind protective walls. City walls symbolized urbanites' desire to keep peasants in the countryside so they could feed the city. From longer distances, foreigners also arrived seeking their fortunes. The Genoese and Lombards went to Sicily, for example, while the Pisans traveled to Sardinia. Migrants arrived from other parts of Europe, too. German artisans and merchants attracted attention in Florence and Venice. Venice also drew most of its domestic servants from a market in the nearby Balkans, while Sicily attracted permanent settlers (Albanians or Greeks) who fled the Balkans as the Ottoman Empire conquered it. Overall, Italy was more a land of immigration than of emigration, and its commercial and industrializing cities were its main attraction.[17]

Italy's powerful merchants valued their city homes highly, and viewed them as the ultimate earthly expressions of human genius.

City dwellers contrasted the civiltà, or civilization of urban life, to the beastly life of peasants in the countryside.[18] Peasants seemed part of the natural vegetation of the countryside, one historian later noted. Even in the more feudal South, where land mattered more than commerce as a source of wealth, the wealthy preferred life in urban courts to isolated castles. This sharp cultural division between urban and rural Italy shaped the peninsula's life and politics into the 1970s, separating the residents of Italy into two peoples who often seemed as different as two races.[19]

Even more than economic power, the concentration of political power in urban hands defined Italy's civiltà. In the republican communes of the center and North of Italy, merchants, industrialists, and skilled workers ruled. Political conflicts were endemic in a land of local loyalties. Italy's northern city states bristled with the towers that symbolized families' prestige and strength; they provided a safe retreat when kin groups engaged in armed conflict with each other. Deadly quarrels broke out between quarters or parishes within a single commune. City governments ended expressions of internal conflict by abolishing (and razing) family towers, and by transforming parish conflicts into games and sporting competitions. Over time, the communal patria consolidated its monopoly over both governance and the use of arms.[20]

Still, political strife remained rancorous and intense. Throughout the late Middle Ages local (Guelf) supporters of the Papacy distrusted local (Ghibelline) supporters of the Holy Roman Emperor, even when neither was intimately involved in imperial politics. Unimportant clients of a Guelf local strongman battled the followers of his Ghibelline counterpart. Violent wars between communes were also the rule. The century-long hatred by Genoese of Venetians originated in economic competition, while Florence, Siena, Pisa, and Arezzo more often battled as they sought control over neighboring territories in Tuscany. Republican governance rarely survived unmodified for long. After 1200, charismatic leaders — called variously despots, dukes, and signori — built dynasties (signorie, principati) that replaced most republican governments.[21]

Political fragmentation and pride in one's city homeland spurred furious competition in cultural creativity, contributing significantly to the development, character, and perfection of civiltà italiana. Like Italy's city dwellers, civiltà italiana had many origins. Italy's merchants contributed the techniques of modern commerce and trade (from double-entry bookkeeping and letters of credit to modes of ship construction). Its Catholic clerics and universities introduced scholarship, law, and administrative theories during, first, Catholic and, later, humanist academic revivals. Love for the city of one's birth, the patria, also associated civiltà italiana with political liberty, especially during the years of the Italian republics. In 1403 Leonardo Bruni, for example, claimed that without justice "there can be no city, nor would Florence even be worthy to be called a city," nor was life in it worth living without liberty. Love of patria remained linked with liberty even in the later writings of Machiavelli.[22]

Scholarly and artistic expressions of civic pride in Italy reached new levels during the Renaissance, albeit under rapidly changing political and economic conditions. After 1450, trade and economic expansion gradually shifted toward the western Mediterranean and then into the Atlantic Ocean; in the process, trade also shifted increasingly out of the hands of Italy's merchants. Central and northern textile production continued vigorous but lost its competitive edge to the manufacturers of England and the Low Countries. Merchants from Italy participated in but did not dominate the coastal Atlantic trade that built the wealth of Antwerp, London, and the Hanseatic cities. Venice's hegemony in the eastern Mediterranean declined precipitously after Constantinople fell in 1453; the subsequent expansion westward of the Islamic Ottoman Empire soon threatened even the Italian mainland. Of Venice's main competitors, only Genoa's mariners and merchants for a time successfully oriented themselves westward. They dominated the trade, banking, and maritime expansion of Aragon and of Catalonia, and somewhat later, of the united Spanish monarchy. Christopher Columbus remains the best reminder of the Genoese explorers and bankers who financed and organized Spain's expansion across the Atlantic and into the Americas. Bankers from Florence too maintained their businesses in England, even after the bankruptcies of the thirteenth century. While still important centers of cultural innovation, Italy's cities no longer organized the international economy.

After 1450, the constant warfare of Italy's many cities and princes, along with their hired mercenaries, proved more than expensive: they prompted repeated foreign intervention by larger and more powerful states forming in northern and western Europe. Beginning in 1490, battles between Bourbon and Habsburg dynasties frequently took place in Italy, as foreign rulers formed alliances with warring Italian princes. The result, often enough, was foreign hegemony over the rulers and merchants of Italy. The bankruptcy of the Spanish Empire in 1627 threatened to end even Genoa's longest-surviving expression of global economic power rooted in Italy. Madrid, Lisbon, London, Amsterdam, and Paris replaced Italy's cities as entrepots. The Atlantic and the Pacific replaced the Mediterranean as the most important arenas of world trade and colonial expansion.[23] By 1800, as the ideas of the Enlightenment swept through northern Europe, visitors found Italy's peasants poor, its cities in varying states of physical decay, its rulers and priests corrupt, and its streets full of mendicants.[24]

Oddly, however, as Fernand Braudel first noted, "the whole sky of Europe" was at first "lit up by these falls" of Italy from its primacy.[25] Italy's "second" ("high" or "late") Renaissance (1350–1550) was a cultural — not a political or economic — rebirth. The creative innovations of the Renaissance extended into the 1600s through the cultural forms of Rome's Catholic Counter-Reformation and the secular music and art of Baroque Venice and Naples. Italy's cities, along with the consolidated seat of Catholicism in Rome, remained culturally vibrant during Italy's economic and political decline. "The magical moment of the Italianization of Europe," as Giuseppe Galasso called it, thus extended, even if gradually fading, almost to the eve of the French Revolution.[26]

Its origins were deep in urban life in Italy. The many institutions of the Catholic church were the first important urban patrons, and much of Italy's early cultural production focused on religious themes. Beginning already after the first millennium, first Italy's wealthy merchants, then its republican communes, and finally their new dynastic rulers (like Florence's Medici family) demanded grand municipal buildings and churches in their hometowns to demonstrate their ascendance. Successful industrialists and merchants competed to make their private residences and places of business equally grand, and to furnish themselves and their homes with the finest clothing, furnishings, and works of art. Merchants, city governments, guilds, and despots became generous patrons of a humbler elite of culture producers. These artists, architects, scholars, composers, and singers created the civiltà italiana of the Renaissance and Baroque. Highly skilled men (and almost all *were* men) built, decorated, and embellished the grand homes, municipal buildings, and churches that became emblems of each patria, and symbols of the patriotism of its residents. They entertained their patrons with madrigals, oratorios, and commedia dell'arte.[27] Tourists today still travel to Italy to enjoy the severity of Florence's civic architecture and the eye-dazzling complexity of Rome's Baroque churches, the frescos of Giotto and Michelangelo, the chiaroscuro of Caravaggio, and the grand canvases of Titian. They still listen to the music compositions of Corelli and Monteverdi.

Italy's cultural creativity after 1500 was the creativity of the periphery. The shift of economic and political power from Italy to the new and imperial nations of western Europe left the country in a geopolitical position little different from the colonies of the New World beyond the Atlantic. Only its colonial masters changed, as Spain and France gradually lost influence over Italy to Austria. In the 1500s, a Spanish official noted contemptuously, "these Italians although they are not Indians have to be treated as such."[28]

By 1700, the political turmoil and economic decline of Italy were having a cultural impact, too. Italy's second Renaissance had extended into a Baroque movement more secular than its equivalent in the rest of Europe. But while new forms of civiltà italiana continued to emerge from Italy in the eighteenth and nineteenth centuries (especially in theatrical and musical forms), the centers of European culture had shifted northward. Amsterdam, Vienna, Dresden, London, and Paris had replaced Florence, Rome, and Venice as the cultural capitals of Europe. The artists and intellectuals of the North, like the young, romantic Goethe, still saw a visit to Italy as an essential stage in their creative development. They admired Italy's past treasures, along with its landscapes and gentle climate ("wo die Zitronen *blühn*" — where the lemon trees blossom), but what inspiration they gained in Italy resulted in cultural expressions in other languages, and other countries.[29]

Even in 1789, after almost 300 years of economic stagnation, the term "Italian" signified something quite concrete and positive in the west. Italian was not yet a noun for a human identity, people or nation. But it was an adjective that described a distinctive range of cultural products — both secular and religious — that the rest of the world found valuable. Throughout Europe and eventually the Americas, too, the politically and economically powerful new nations, and New World economies, continued to demonstrate their successes by acquiring Italian culture, and making it part of their lives. As in the time of Italy's primacy, too, they often purchased civiltà italiana directly from its producers — the migrants of the first Italian diasporas.

The world is my country

Travelers from Italy rarely went unnoticed. Echoing the earlier insights of the vernacular poet, a popular saying of the early modern era noted, "In whatever quarter of the world — you open an egg and a Genoese will spring from it."[30] Migration had characterized life on the Italian peninsula since humans first inhabited it. Examining the Middle Ages, Renaissance, and early modern eras, demographers of Italy now distinguish between those who traveled long distances (beyond the sea (oltremare) or beyond the mountains (oltremontane)) and those who traveled shorter distances. The latter dwarfed the former in numbers.[31]

It was the relatively small group of long-distance migrants who did most to "Italianize" the rest of the world. Peasants, shepherds, artisans, and smaller businessmen looked for seasonal agricultural work nearby, moving from the mountains to coastal areas, or between the peninsula and its nearest island of Sicily. They left Italy's Alps and Apennines to find work in the Po Valley or in the valleys of Switzerland and France and the plains of the French littoral. By contrast, Italy's highly skilled and educated artists and scholars, along with its merchants, were more likely to undertake long journeys abroad.

The migrations that spread civiltà italiana beyond Italy made the country's early cultural elite a cosmopolitan group.[32] Their familiarity with life and work abroad made their cosmopolitanism even more striking as Italy fell into economic and political marginality. Many of the names of Italy's earliest migrants beyond the mountains and the seas are well known to students of European civilization; they include Marco Polo, Columbus, Leonardo Da Vinci, Casanova, Titian, and Napoleon Bonaparte's Tuscan ancestors.

With no modern nation states to record their comings and goings, the earliest long-distance migrants of Italy easily disappear from history. Yet precisely because they were an extraordinary and relatively small group, they have attracted their catalogers. The proletarian mass migrations and a new nationalist movement in Italy in the 1890s inspired a first round of biographical dictionaries on Italy's earliest migrants.[33] Somewhat later, fascists' obsessions with Italy's former greatness encouraged Ugo E. Imperatori to prepare his own biographical sketches of migrants, drawing on the earlier works.[34] An expanded version of his research appeared in 1956, and in its preface Imperatori urged "every city that is proud to have given birth to some of the meritorious" men to purchase a copy. In it, Imperatori provided short biographies for 3200 migrants. The largest group comprised men and women who migrated in the nineteenth and twentieth centuries. Still, a not negligible number of over 1400 entries recorded the births (place and date), hometowns, occupations, destinations, major accomplishments, and deaths (place and date) of those leaving Italy before 1789.[35]

Imperatori listed rather small numbers of migrants for the thirteenth through the fifteenth centuries; his entries for subsequent centuries number in the hundreds. Of course, even the latter numbers seem trivial when compared to those of the mass migrations from Italy after 1870. Scholars, too, agree that Italy's early long-distance migrations were not particularly numerous by either ancient or modern standards (unless one counts the 60,000 or so mercenaries recruited in Italy to fight in Flanders and Spain in the seventeenth century). They were not the equivalent of the *Völkerwanderung* that disrupted the Roman Empire. The expulsion of Moors and Jews from Spain in the 1490s, and the flight of Huguenots from France in the seventeenth century, were numerically far more significant.[36] Of course, Imperatori's brief biographies serve to outline only the destinations and structures of Italy's migrations, not their volume.

The lives of the 1400 elite migrants Imperatori describes suggest a significant transformation in migration patterns that accompanied Italy's transition from late medieval economic primacy to economic decline amidst the cultural innovation of the Renaissance and Counter-Reformation. The diasporas formed by migrants differed in the two periods. In Imperatori's brief biographies, transalpine and seaward migrations from Italy remained in surprisingly stable balance over many centuries, as Table 1.1 shows. However, migrants' exact destinations shifted after 1500. Civiltà italiana defined a European (or western) civilization largely because the Ottoman Empire terminated the earlier expansion of the merchants of Venice and Genoa into the Balkans and the eastern Mediterranean after 1500.[37] Thereafter, civiltà italiana traveled most consistently across the Alps to central, western, and northern Europe, with France gradually losing place to Austria and Germany as the main destination of Italy's migrants. Beyond the seas, Italy's migrants had first moved in large numbers into the eastern Mediterranean and (through the lands of the adjoining "Orient") into Asia. After 1500, their migrations shifted toward the western Mediterranean, and from there, into the Atlantic world. Transalpine migrations seemed about to replace seaward migrations after 1700, but migrations to the Americas (see Chapter 2) then quickly took off, returning oltremare/oltremontane migrations to their long-term balance.

The status and occupations of migrants also changed as Italy fell from its economic primacy as trade crossroads in the Mediterranean to become primarily an exporter of culture. Before 1500 scholars and

Table 1.1 Selected destinations of migrants from Italy, 1200–1790 (%)

	1200s	*1300s*	*1400s*	*1500s*	*1600s*	*1700s*
Far East	12	10	5	8	11	4
Mediterranean	25	20	34	22	22	16
Atlantic Islands/ Americas	–	3	2	4	4	3
Total oltremare	37	33	41	34	37	23
France	39	35	19	24	20	17
Germany/Austria	7	10	8	13	23	27
Britain	7	8	5	6	5	14
Central/East Europe	7	8	13	11	8	10
Total transmontane	60	61	45	54	56	68
(Number, oltremare and transmontane	41	59	148	386	332	477)

Source: Ugo E. Imperatori, *Dizionario di italiani all'estero (dal secolo XIII sino ad oggi)* (Genoa: L'Emigrante, 1956).

writers together formed the largest single occupational category of the small group of elite migrants (155) described by Imperatori. Missionaries and merchants were the second and third largest categories. As Italy's economic power declined in the 1400s, mariners and explorers like Columbus and Cabot, in service to the Spanish, Portuguese, and British empires, replaced merchants as second or third largest categories. After 1500, artists and humbler culture producers became the largest category among the much larger group of migrants Imperatori includes (608). The visual artists and architects of the 1500s and 1600s (including men like Antonello di Messina and Giovanni da Verona) gradually gave way to singers and musicians (Pietro Nardini), musical composers (Baldassare Galuppi), and actors (Cesare D'Arbes) in the 1700s.[38] During the Enlightenment, very few visitors detected much change or progress in Italy. But already in the 1500s, Italy had also begun to export scientists and technicians, like Francesco Antonelli from Ascoli Piceno who specialized in the construction of military fortifications and found work with Ferdinand II in Landsberg and Hungary.

Italy's migrants enjoyed very differing relations to the societies that received them during and after their homeland's economic decline, and they thus formed rather different networks of communications and different types of diasporas. The earliest migrants — whether clerics or merchants — tended to form self-governing communities, and to live apart from the foreign cities where they sought markets or converts, while maintaining good contacts with their counterparts in other foreign locations. Later, from the Renaissance until the French Revolution, humbler producers of civiltà italiana, along with diplomats and scholars, most often traveled as individuals following offers of employment and patronage. Their strongest ties were to their patrons and to their homeland, not to migrants from Italy living and working in other places around the world.

The earliest migrants from Italy formed three kinds of settlements. In the late Middle Ages, and mainly in the eastern Mediterranean, Venetian merchants were colonizers in the modern sense, and they administered their Aegean islands and populations as colonies under Venetian law.[39] They brought their homelands with them as they traveled, and imposed its laws and mores on the lands and peoples they occupied. Elsewhere, around the Mediterranean and the Black Sea, and in transalpine Europe as well, early merchants from Italy's cities instead formed merchant colonies as part of a trading diaspora. Here they governed only themselves, but under the laws of their homeland patria, not of the city where they resided. These migrants too — as the vernacular poet also noted — brought their homelands with them as they traveled, but they created little Venices and other Genoas rather than conquering and ruling new lands and peoples.

Italy's trading diaspora had its Catholic counterpart, too. In the 1200s, Catholic missionary zeal sent many Franciscan and Dominican migrants from Italy to the eastern Mediterranean, Asia, and Africa. Later, Jesuits like Rodolfo Acquaviva (who went to Goa) and Giovanni Antonio Cantova (who served in missions in Mexico, the Philippines, and the Caroline Islands) spread Catholic influence to the Americas and Asia. Merchant, mission, and colonizing diasporas were of greatest importance before 1500; thereafter, only Catholic religious orders created self-governing, and linked, communities, mainly in the French, Spanish, and Portuguese empires.

At the centers of the Venetian colonial empire, the trading diasporas of Genoese and Lombards, and the Catholic diaspora were particular sending cities in Italy — Venice, Genoa, Milan, Rome. In each case, however, migrants in one location usually enjoyed relatively good contacts with fellow merchants or clerics in other places, and traveled among them, not just back and forth between the patria hometown and little Genoas abroad. Merchant families expanded their businesses by sending brothers or cousins to settle and trade in a wide variety of locations.[40] A man like Francesco Pegolotti Balducci, from a merchant family of Florence, had worked in Antwerp, London, and Cyprus by the time he was an adult; he later traveled to Cyprus as a representative for a Florentine bank. His Genoa counterpart, Antonio Malfante (who had first traded in Spain in the early 1400s), later explored commercial possibilities for his company of merchants, going into sub-Saharan Africa as far as the mouth of the Niger River. Similarly, in the Catholic diaspora, the Venice-born Marco Barbo, the Patriarch of Aquilea in the 1400s, went later to Hungary to mediate a conflict between Hungarian and Polish claimants to the throne of Bohemia. Both the religious and commercial purposes of these diasporas, as well as the imperial expansion of Venice, required ongoing connections among far-flung locations.

Although highly skilled and relatively well educated, the men who left Italy after 1500 differed significantly from the Catholic and merchant colonizers of the late Middle Ages, and so did the diasporas they formed. Migrants like the architect Domenico Martinelli (born in Lucca in 1650), the singer Gaetano Maiorano (born in Bari in 1703), or the painter Andrea Procaccini (born in Rome in 1671) were part of Italy's elite but they were nevertheless labor migrants, not merchants. Unlike the earlier migrants, who with very few exceptions returned home again to prosperous and important lives in Italy, migrants after 1500 were more dependent upon foreign employment (see Table 1.2). While they, too, retained ties to their home cities, their rates of return to Italy dropped to a level comparable to modern labor migrants (e.g. somewhat greater than 50 percent).

Like later migrants, too, the creators of Renaissance and Baroque civiltà italiana depended on patrons and foreign employment. Under his theatrical name "Caffarelli," Maiorano sang in all the

Table 1.2 Elite migrants who returned to Italy, 1200–1790

	Percentage returning	*Number*
1200–99	71	28
1300–99	62	37
1400–99	79	91
1500–99	62	242
1600–99	55	217
1700–89	58	165

Source: Elite migrants only; Imperatori, *Dizionario di italiani all'estero.*

major theaters of Europe, and enjoyed special success in London and Paris. Martinelli was invited to Austria in 1690 to design the Viennese palace of the Prince of Lichtenstein; he later built gardens for the Count of Kaunitz in Austerlitz. Leopold I then sent him to Warsaw, Prague, Moravia, and Prussia to work on various building projects, a patronage which discouraged Martinelli from accepting an invitation from the King of England to plan buildings there. During the same years, Procaccini enjoyed the patronage of Spain's Philip II, whose portrait he painted before undertaking further canvases for the Royal Palace in Madrid.

As employees of wealthy patrons abroad, visual and performing artists could usually expect quick incorporation into the economically powerful new urban centers of transalpine Europe. Migrants with powerful patrons did not form the self-governing communities typical of the merchants and missionaries. Their closest ties were not to other communities of painters, architects, or musicians from Italy who lived elsewhere in a diaspora, but instead to their patrons and to their birth cities, to which many still returned.

As Italy declined, ever more of its cities specialized in producing particular types of exportable culture. In doing so they encouraged the formation of diasporas that linked particular cities in Italy to particular locations abroad, rather than encouraging ties among migrants in various foreign locations. Even before 1500, explorers and merchants most often left from Genoa, Venice, and Florence. Scholars, writers, jurists, and lawyers left university towns like Bologna and Padua. After 1500, Florence and Rome sent abroad artists and architects while singers, musicians, and actors more often originated in Rome and Naples. In later centuries, too, Italian towns would develop a wide range of far humbler, but equally portable crafts and trades, and migrants would leave home in search of patrons for their wares and services.[41]

Collectively, the Italian cities producing civiltà italiana defined the cultural heart of the geographic expression that was Italy. The Italy of civiltà italiana was undisputedly urban. Over time the migrants leaving Italy's six largest cities (Venice, Genoa, Milan, Florence, Rome, and Naples) slowly declined from two-thirds to about one-third of all migrants. In the thirteenth century, the urban homeland of civiltà italiana included only the largest cities of the northern and central parts of the peninsula. But over time the smaller cities of the North and center and the largest cities of Naples, the South, and Sicily also became part of the Italy that produced and exported civiltà italiana.[42]

The fact that migratory urban elites created civiltà italiana, and exported it abroad, made them and their culture as cosmopolitan and European as it was Italian. Although nationalists of the nineteenth century later claimed that these early migrations created the first Italians, as well as Italian culture, they exaggerated.[43] The loyalties of the migratory merchants remained strictly limited to those of their own families and patria. The artists who created and exported civiltà italiana felt no apparent solidarity with other occupants of Italy, nor did they live among them when they migrated. Like Dante — himself an exile from Florence — the migrants of Italy's early diasporas could have claimed instead that "the world is my country." They were more interested in the world beyond the mountains and seas and in their birthplace cities than they were in Italy or other Italians.

Patria e natio

The men who created civiltà italiana did not consider themselves Italians. Other forms of identity — from the particular ties of kinship, to a universalizing religious faith, to the urban birthplace (patria) or its surrounding region — remained of greater salience. Use of the term natio did emerge in the first of Italy's diasporas, however, as migrants confronted strangers who knew little of the differences among them. Even abroad, a natio meant a specific occupational group with specific origins in one patria; it did not include every migrant from anywhere in Italy. The expansion of a few city states into larger, regional dynastic states in Renaissance Italy gradually strengthened regional loyalties at home. But only in very exceptional cases did migrants during this era see themselves, or come to be seen, as Italians.

The first nations of Italy developed among Italy's migrant merchants and students in the late Middle Ages. Merchants abroad sometimes called themselves parishes or guilds. But, as an early historian of merchants in the Low Countries of northern Europe wrote,

> As soon as the Italians established themselves in Flanders, they began to organize "nations" or colonies which were composed of all the merchants from the same city. . . . Genoese, Venetian, Lucchese, Florentine . . . Milanese nations. . . . One of the first tasks of the newly formed nations was to secure official recognition of their

> incorporation and to obtain commercial privileges from the local authorities.[44]

A merchant natio generally enjoyed carefully negotiated contractual relations to the rulers of the port cities where they built their residential and business quarters (called "fondaci").[45] Most fondaci in turn governed themselves under the rules of their republican hometowns. They were extensions of the patria on foreign soil, defined not by their citizens' place of residence but by their place of birth.

Just as the arrival of cosmopolitan migrants fixed the location of a place called Italy in the minds of Europeans, so too foreigners developed notions of the cultural character of Italy's residents when confronting the migrants. During the early years of Italy's economic primacy, natives sometimes viewed these merchants from Italy in terms we today associate with anti-semitism — as conniving possessors of financial secrets that gave them undeserved wealth and power over poorer or less sophisticated natives. Xenophobic violence against Italy's merchants was frequent in the Mediterranean, especially where local populations were Muslim or Greek (and Orthodox). On more than one occasion, natives attacked and burned the Black Sea fondaci of Venetians and Genoese; Arab and Turkish sultans regularly revoked the special privileges of Italy's merchants, and expelled them from their port cities. Both Englishmen (in the 1200s) and Parisians somewhat later complained bitterly about the influence of foreigners from Italy in their markets and courts.[46]

With the end of Italy's economic primacy, European reactions to the arrival of newer migrants from Italy became more favorable. Europeans saw the Catholic clerics, artists, and musicians of Italy as religious, well-educated, gifted, and proud people, specially endowed by their birth at the center of the classical world, of Christendom, and of the Renaissance.[47] The labor migrants and clerics did not hold themselves apart, but mingled more extensively with natives in the courts and universities where they found employment. Even as Italy declined into economic and political dependency, foreign estimation of its cultural émigrés seemed to rise.

Rarely did Europeans praise or heap scorn on "Italians," however. Even abroad, praise or criticism descended on the heads of migrants from a particular patria or region — from Genoa or Lombardy, Tuscany or Rome. The fact that a fondaco of merchants organized itself as a communal settlement but called itself a natio reminds us how narrowly men of the late Middle Ages and Renaissance defined loyalties. It suggests as well that the term natio — and with it concepts of nationality — emerged from political and economic interactions, and from legal treaties and trade — not from civiltà italiana in its artistic, scholarly, or musical expressions. Those cultural forms instead encouraged close ties between native-born patrons and their migrant employees from Italy. The artists, diplomats, or musicians of the Renaissance never claimed the privileges or obligations of self-government or a natio for themselves.

In northern Europe from an early date, natives ignored or collapsed communal differences among Italy's migrants into regional ones. Indeed, regional identities seem to have emerged abroad well before they developed in much of Italy. At the important early trading fairs in Champagne, French merchants praised, or more often complained of, the Tuscans (a group that included the warring peoples of Florence, Pisa, and Arezzo). They spoke, also, of the Lombards (by which they meant both merchants of Venice and the moneychangers from Milan).[48] Meanwhile, at the University of Paris in the 1200s, students from Italy were counted instead among the nation of "latini" (speakers of Romance languages), and gave their name to their residential quarter on the Left Bank of the Seine.[49] Only in exceptional cases did natives view early migrants as Italians. Marco Polo was a Venetian migrant to China's Kubla Khan in the 1200s, but in the sixteenth century, in Macao, Chinese merchants instead distinguished between traders and mariners from Portugal and religious missionaries who hailed from "I-tal-i."[50] In Europe, the only thing that the many migrants from Italy seemed to share was their Catholicism, but this scarcely differentiated them from natives before the Protestant Reformation.

Life abroad encouraged increased interdependence and identification across communal lines, fostering the development of regional nations. Certainly, historian Armando Sapori believed this to be true of Italy's merchants.

> If at home he was quarrelsome, suspicious, and cold even in his relations with his family, when he was away from Italy he forgot any political rancor and drew closer to his fellow citizens, inspired by a sense of solidarity, which was based on both practical and idealistic considerations.[51]

Still, solidarity never extended to all migrants from Italy. Abroad, as at home, scholars continue to admit, "there were no Italians."[52]

At least until the Italian Renaissance, kin, feudal, and factional identities (Ghibelline; Guelf) and communal identities based on love of and pride in the communal patria remained of greatest significance on the Italian peninsula itself. This is scarcely surprising. A republican city state automatically awarded citizenship (cittadinanza) to residents (cittadini) of the city (città) but not to those born in the countryside immediately surrounding a town. They demanded political participation and military service of their male citizens while debating continuously the conditions under which foreigners — from the countryside, the next town, or region, or from farther afield — could naturalize and obtain the rights and duties of citizens. Their desire to see their patria prosper and grow conflicted with their fear of foreigners (forestieri or outsiders). Constant battles with nearby towns made any foreigner a potentially disloyal source of disorder, even after naturalization.[53]

As republican city states gave way to rule by despots and to larger principalities and dynasties after 1400, regional loyalties grew in importance. As early as the 1200s, a small group of cities of the North had experimented temporarily with a military alliance they called the Lega lombarda (the Lombard League). Linguistic similarities may have been more important than politics in defining regional loyalties. An occasional intellectual like Petrarch or Dante spoke of Italian as a literary language. These men wrote in their vulgate Italian dialects, which departed significantly from classical, and even clerical, Latin. Dante's Tuscan dialect did ultimately define the modern, national language of Italian, but only much later. In Dante's age, mainly Tuscans spoke or understood it. Even in a linguistic sense, an early historian of the Italian nationalist movement, Giuseppe Baretti, noted that the inhabitants of Renaissance Italy "lived as if they occupied different islands," each speaking its own regional dialect — Piedmontese, Tuscan, Neapolitan, Roman, or Sicilian.[54] The differences in these dialects were not trivial; as late as 1900, a folklorist discovered in Italy seventeen mutually unintelligible words for "little" — ranging from zicu and zinnu to cit, cico, minore, and piccirillu.[55]

Mutual distrust persisted on the Italian peninsula, although it too increasingly found regional expression. Baretti continued his analysis by emphasizing that

> the Genoese do not like the Tuscans, and the Tuscans show no favor for the Venetians or the Romans, and all of the

> nations, without really knowing why, were animated by the most ridiculous antipathy toward one another.[56]

The Genoa and Venice of which Baretti wrote had, since 1200, expanded from city states into regional republics with extensive inland and island territories. Both Tuscany (under a despot, and later a Grand Duke) and Rome (under the Pope) had also become dynastic states encompassing expanded regions that contained formerly independent city states. Baretti's description erred in only one important respect. The regional groups he described almost never referred to themselves as nations.

Perhaps this was because state and region in Renaissance and early modern Italy were rarely coterminous. Most people sharing a regional identity had no state of their own in 1789, nor did they often demand one. Marchegiani found themselves as uncomfortably ruled by the Pope in Rome as did Bolognesi or Umbrians to the north and northwest. Tuscans, Modenese, and Milanese were ruled by dynasts of their own regional background at times; at other times they found themselves dominated by French- or German-speaking princes. The Bourbon rulers of the southern provinces of Italy had the double disadvantage of being foreigners who ruled a country of mutually distrustful Sicilians, Basilicati, Neapolitans, and Calabrians. None of Italy's many regions generated a national dynasty or dynastic state resembling France or England. Italians, when they existed at all, lived mainly in the diaspora. And even there they existed mainly in the minds of the English, French, Spanish, and Turkish who employed or traded with them.

The legacy of civiltà italiana

Devotion to a patria or a region in Italy was not the equivalent of modern nationalism; these loyalties developed more from love of a place than identity with a people defined through shared biological descent. Italy's earliest nations formed abroad through commerce, trade, and political negotiation, not through the creation of civiltà italiana or the tug of religious faith. It is thus surprising that civiltà italiana later became so important to Italian nationalists eager to create an Italian national identity and to organize a nationalist political movement to create an Italian nation state.

Civiltà italiana was an odd foundation for any of these later developments. A culture of inclusion, cosmopolitanism, and incorporation, civiltà italiana had proved easily transportable across parochial boundaries of states, rulers, languages, and regions. Civiltà italiana was never rooted definitively in any one particular patria; nor was it the creation of a single natio. Its creators came from many cities and regions in Italy. When they migrated abroad, furthermore, the producers of civiltà italiana increasingly lived among the native-born elite, seeking, and usually enjoying, good access to native religious and dynastic centers of cultural influence. Finally, civiltà italiana, as a culture, neither acknowledged nor spoke to the fundamental chasm of class. It was not a product of the countryside, nor did it represent the lives of the peasants who formed the majority of Italy's growing population. In the countryside, people neither produced nor knew of civiltà italiana; wealthy, educated foreigners knew it better than they did.

The sharp divisions of economic class and of city from countryside left no mark on civiltà italiana, but they shaped both the creation of an independent Italy and all of Italy's subsequent diasporas. Italy's nationalists came to believe that a nation of Italians had created civiltà italiana, and that Italy needed its own state in order to revive the civilizing mission of civiltà italiana. While some viewed the growing migrations of the nineteenth century as symbols of moral and intellectual collapse, others saw in them new opportunities to spread Italy's influence abroad. A few dreamed of Italians colonizing the world in the same way that Venetian merchants of the 1200s had colonized the eastern Mediterranean. Barring that, nationalists hoped at least to create colonies of Italian influence comparable to those of Genoese merchants and Catholic missionaries of the late Middle Ages. The shadow of civiltà italiana hung heavily over Italy's evolution into a modern nation state and over that state's evolving relationship with its new diasporas of migrant proletarians.

2

MAKING ITALIANS AT HOME AND ABROAD, 1790–1893

"What do you mean by a nation, Mr. Minister? Is it the throng of the unhappy?"[1]

In the years after the French Revolution, mass migrations of ordinary workers and peasants traveling beyond the seas and mountains developed alongside a political movement for a unified and independent state of Italy. Few at the time saw the Risorgimento and migration as connected, yet both were rebellions against the status quo of pressing poverty in the countryside, clerical and aristocratic privileges in the cities, and foreign interference everywhere. Both groups of rebels linked their hopes to Italy's lost influence, and to its earliest diasporas. Nationalists saw the Risorgimento as a modern, political expression of civiltà italiana. Meanwhile, humble migrants who displayed dancing bears, cleaned chimneys, or shined shoes nodded to their predecessors by calling themselves — the practitioners of far humbler trades — migratory "artists."[2] Neither Italy's creation nor the movement to define a new nation of Italians can be understood apart from the country's migrations and many diasporas.

Choosing a watery metaphor, Italy's nationalists called their movement for national independence a "Risorgimento," or a bubbling forth of the creative powers from the deep spring of an Italian nation.[3] That nation had not yet recognized itself. Italy's new nationalists struggled to counter the local and regional loyalties of earlier nations and to build a modern nation embracing residents throughout Italy. Those earlier nations of Lombards and Tuscans had first emerged abroad, in the diasporas formed by Italy's merchants. The modern Italian nation, too, often seemed to find form more easily outside of Italy than within it.

Migration and nationalism were closely entwined in the lives of the activists of the Risorgimento. One in three Italian nationalists migrated abroad, for their activism often forced them, as they dramatically put it, onto "la via d'esilio" (the path of exile).[4] Migration also became a link between segments of Italy's political elite and the peninsula's humbler residents. The adversarial relations of Italy's two races — the rural poor and the urban bourgeoisie — was the largest single challenge facing Italy's nationalists, even those of the liberal aristocracy. Nationalists who fled into exile often saw labor migrants more as part of an Italian nation than their contemporaries who remained at home. They also sought more insistently and with greater success to convert migrant workers to their nationalist cause, and make them part of the nation of Italians. Exiles like Giuseppe Mazzini and Giuseppe Garibaldi worked with labor migrants to create forms of diaspora nationalism powerful enough to shape events at home.[5] A cross-class collaboration, diaspora nationalism departed sharply from that of Italy's leaders, who remained hostile to plebeian activism. As a result diaspora nationalists more often became critics than governors of the new nation state of Italy.

Italy's middle-class nationalists, not a modern Italian nation, created the independent state of Italy in 1861. It was a nation state in name alone; only 2 percent of the citizens of Italy voted in the plebiscite creating it. After 1881, about 6 percent of Italian citizens, including a very few ordinary workers, could vote; few of the other 94 percent felt much loyalty to a nation of Italians. The main tie between Italy's government and Italy's citizens was through universal male military service. Italy's government remained distrusted by most of those it sought to govern; its diasporas evolved into global webs connected largely to an individual patria or paese rather than to each other or to the Italian nation. Excluded from the nation and political life at home, the poorest Italians by the 1890s seemed poised on a threshold. Would they remain the racially inferior criminals their rulers too often believed them to be? Or would they instead become the most visible of Karl Marx's workers of the world? Or something else entirely?

The Risorgimento and the two races of Italy

In some ways, it is surprising that Italy's new nationalists troubled themselves at all over Italy's poor or their relationship to the Italian nation. The Risorgimento began as a cultural movement inspired by civiltà italiana; nationalists argued that the unique genius of an

Italian nation had created civiltà italiana before falling to foreign tyranny. The origins of the Risorgimento and its concept of the Italian nation were even more thoroughly bourgeois and secular than civiltà italiana, for even Catholic thinkers soon distanced themselves from its ideas. Remaining outsiders, the much larger group of Italy's peasants, along with the landless poor, nevertheless shaped the course of the Risorgimento with their own expressions of discontent. Most bourgeois nationalists concluded that plebeians ("la plebe," they called them — emphasizing their lower-class origins — not "il popolo," a more flattering and inclusive term, "the people") needed police surveillance, not democracy. Ultimately, that is what an independent Italy delivered.

Nationalists more easily overcame Italy's regionalism than its class differences.[6] Under the influence of French Enlightenment ideals, the southern students of Giambattista Vico abandoned Latin to converse and lecture in the local vernacular. In the North, historian Ludovico Antonio Muratori published collections of medieval documents to establish that a common Italian cultural heritage stretched back to Italy's primacy. By 1800, a poet like Ugo Foscolo — born on the Ionian island of Zante to a Venetian father and a Greek mother — felt himself proudly Italian; so proudly Italian, in fact, that Foscolo used the metaphorical language of romantic fiction to argue that true Italians (like him) would commit suicide rather than fall again to foreign invaders. Death would guarantee that "their bones at least rested on the land of their fathers."[7]

Foscolo wrote his suicidal fantasies of patrilineal nationalism during Napoleon's battles with Austria for control of northern Italy. By then, cultural nationalism had already given way to agitation for political change and become associated with anti-clericalism, as it was, too, in France. The French Revolution, followed by Napoleon's invasions of Italy, sparked the erection of liberty trees and the proclamation of republics North and South, along with calls for national unity under an Italian "tricolore" (the colors red, green, and white). Calling themselves patriots (because they loved their patria), activists had considerably enlarged their understanding of patria to embrace the entire Italian peninsula. Many were also republicans who linked love of patria to love of liberty. Italy's first modern republican revolts (in 1799) quickly fell to Austrian and Russian armies in the North, and to a religiously motivated and led peasant movement, the Sanfedesti ("The Most Christian Armada of the Holy Faith"), in the South. Napoleon's subsequent conquest of Italy, and his Italian monarchy, imposed a modicum of unity on the deeply divided North, while in the South Napoleon's brother ruled.[8]

The ideas of independence and liberty persisted in local secret societies after the restoration of the Bourbon, Savoyard, Austrian, and papal ancien régimes in 1815. Isolated, regional, revolutions by supporters of constitutional government followed in Turin, Messina, and Naples in 1820–1. The Carboneria conspirators flourished further south. Central Italy saw its own revolts against papal authority in the early 1830s. By 1848 urban revolts, provisional republican governments, violent food riots and land occupations, and bloody tax revolts occurred everywhere from Palermo to Venice — as indeed they did throughout Europe during its revolutionary spring of that year.[9]

After 1848, Italy's many movements for political change coalesced into a movement for national unification, bridging the many regions of the peninsula. Catholicism — which transcended Italy's regional differences — did not provide the bridge among them. For a brief moment, a few nationalists had hoped a liberal Pope might help create a unified Italy. After 1848 the distance between nationalists and clerics broadened into rigid hostility, and moderates increasingly pinned their hopes on the King of Sardinia, who had responded positively to liberal demands in 1848, as a potential leader of a united nation.[10] Anti-clericalism proved a stronger link between nationalists of differing regional backgrounds and political dreams.

First Florence, then Turin, became popular places of refuge for activists in failed revolts elsewhere in Italy. In both cities, refugees could expect to find sympathy and financial aid from native nationalists. Even southerners opposing the Bourbon monarchy now traveled the long path of exile northward. Thus, Paolo Imbriani (a civil servant in the Neapolitan Republic in 1848) fled to Tuscany, and taught law at the university in Pisa until he could return home in 1860. Pietro Aristea Romeo, an engineer from Reggio Calabria, fled after Bourbon soldiers ordered him to parade the head of his beheaded uncle on a pike through the streets of Naples after a failed rebellion. Together with his father and other family members, he found refuge in Tuscany, Turin, and Paris.[11] Without leaving Italy, exiles behaved much like other migrants, and founded mutual aid societies like the Associazione dei Profughi Italiani (Association of Italian Refugees) and the Società dell'Emigrazione Italiana (Italian Emigration Society) in Turin.[12]

Italy's Risorgimento nationalists faced a problem even larger than regionalism in the divide separating Italy's "two races." Middle-class

Table 2.1 Social class composition of Risorgimento and other European nationalist movements (%)

	*Italy**	*Finland*†	*Bulgaria*†
UPPER		5–12	0
Aristocracy	7		
High bourgeoisie	5		
High military	6		
MIDDLE			
Clerics	4	20–30	10
Professions	31	32–55	18
Intellectuals	16		
Low military	15		
Low bourgeoisie	6		
LOWER			
Artisans	5	1–3	15
Plebeians	2	1	2
Peasants	1	2–5	33

Sources:
* Michele Rosi, *Dizionario del Risorgimento nazionale: Dalle origini a Roma capitale; Fatti e persone* (Milan: Casa Dottor. Francesco Vallardi, 1930–7), and Francesco Ercole, *Il Risorgimento italiano* (Milan: B.C. Tosi, 1936).
† Miroslav Hroch, *Social Preconditions of National Revival: A Comparative Analysis of the Social Composition of Patriotic Groups among the Smaller European Nations* (Cambridge: Cambridge University Press, 1985).

men dominated the Risorgimento in Italy, as they did nationalist movements everywhere in nineteenth-century Europe (see Table 2.1).[13] Civiltà italiana's influence on national sentiments revealed itself in the large number of nationalist intellectuals; amazingly, almost 5 percent of Italian nationalists were poets and song writers. Bourgeois critiques of the many ancien régimes of Italy also fostered nationalism. In Italy's center and North, intellectuals and professionals sharply criticized dynastic rulers whose archaic taxes and laws stymied industrial investment and agrarian reform.[14] Professionals and lower-ranking soldiers may also have resented their nepotism in filling government positions with fellow aristocrats.[15] In the South, critics of the Bourbon dynasty in Naples more often raised democratic than economist critiques of the monarchy.

Notable for their absence among nationalists were women, clerics, and the vast majority of Italy's population — its peasants and poor. A mere 1 percent of Italy's nationalists, women showed little outward interest in the movement — although this reflected more their exclusion from public life than any advantages they enjoyed under ancien régimes. Women nationalists like Lucia De Thomasis were wives of bourgeois, urban patriots. They became hostesses of nationalist salons, intellectual muses to prominent men, and nurses of wounded family members; occasionally they emerged in public as petitioners on behalf of imprisoned male relatives. When women of the laboring classes became activists they too joined husbands, as did Teresa Scardi who fought on the barricades in Forli in 1831 and then took the path of exile.[16]

Active patriots in at least a few European countries in the nineteenth century, the clergy constituted a tiny group among Italy's nationalists. Hostile to liberalism and appalled at the Roman revolution of 1848, clerics and humble Catholics alike viewed the virulent anti-clericalism of the bourgeois nationalists with horror.

Peasants in Italy were an invisible 1 percent of Italian nationalists; they had none of the importance in this national movement that they would have in central and eastern Europe later in the century. Artisans and petty craftsmen — more likely to live in cities, to know how to read and write, and to aspire to upward mobility into the middle classes — found more to like in the nationalist cause. Overall, however, nationalist sentiments were not widespread among Italy's poorer residents. The poor of Italy, including the women among them, had other ways to express their discontent.

Especially in the South, national ideals flourished among some educated and bourgeois Italians because they feared the rebelliousness of the poor and wanted a national state to protect them, and their property, from it. Middle-class Italians everywhere complained of high levels of disorder under the ancien régimes. They noted recurring revolts of peasants (who burned their landlords' houses, attacked tax collectors, and murdered hated employers) and food riots of hungry urban women trying to procure sufficient food for their families.[17] They feared the diseased, and often almost naked, vagrants who wandered the countryside, and into Italian cities, searching for work and alms. They helped create religious and municipal systems of surveillance to control mendicants and the poor women who fled their home villages to bear children out of wedlock. Such fears are understandable. In the Po Valley town of Lodi in 1830, for example, 25 percent of the population officially registered as paupers. To the horrified bourgeoisie, they were all potential criminals — brigands and thieves, they called them.[18]

Thus a nationalist like Ugo Foscolo saw no room for Italy's poor in the Italian nation. He did not see disorder — as historian Eric Hobsbawm later would — as political action.[19] "There is no case for discussing the plebes," Foscolo concluded, "since in any form of government all they need is a plough or some other means to earn their bread, a priest, and an executioner." Worse, he noted, "when they do move it always finished in theft, blood, and crime."[20] Moderate nationalists like Foscolo wanted a unified Italy with a strong state, a modernizing economy, and public order — not democracy. Even republicans among them often blamed the collapse of Naples's republican revolution in 1799 on peasants who blindly followed the Bourbon counter-revolutionary priest Fabrizio Ruffo and his Sanfedesti movement.

By 1840, however, more democratic nationalists had begun to challenge the moderates' definition of the Italian nation. Some, like Neapolitan lawyer Gerardo Mazziotti, were ex-Jacobins and anti-clerical republicans who — inspired by the French Revolution — organized conspiratorial movements (called Carboneria) to overthrow the Pope or aristocratic rulers after Napoleon's demise.[21] More influential, however, was the much younger exiled republican Giuseppe Mazzini, who in 1835 denounced the arid ritualism of Italy's secret societies, and quickly became the most eloquent spokesman of a new generation of nationalist, radical republicans. Although vaguely anti-clerical, Mazzini's nationalism proved complex, abstractly spiritual, and a product of his universalist hopes for the liberation of all human spirit along national lines. Mazzini organized not only a Young Italy movement; he saw it linked to his other creation, Young Europe. Mazzini's was a voluntary and emancipatory nationalism that grew out of love of la patria; his nation was not a biological or descent group.[22] Mazzini gave some place to the aspirations of the urban poor to escape oppression through nationhood and — more importantly — through armed struggle. Much like later anti-imperialists in the Third World, the otherwise personally peaceful and ascetic Mazzini argued that "Insurrection — by means of guerrilla bands — is the true method of warfare for all nations desirous of emancipating themselves from a foreign yoke."[23] Insurrection was exactly what his moderate opponents feared.

Mazzini expected that the educated would lead poor men into insurrection to create a united and independent Italy; in his nation, bourgeois men acted and directed while peasants and women accepted their guidance and followed their orders. (At most, Mazzini seemed to imagine women raising republican and nationalist children — as had Mazzini's mistress, Giuditta Sidoli.)[24] Mazzini and his bourgeois followers planned many insurrections, usually while in exile elsewhere in Europe. None had popular support; all, including that of the Bandiera brothers in Calabria in 1844, failed. Still, Mazzini's followers challenged the moderates by imagining a nation that included common men, at least as soldiers.

By 1850, even smaller numbers of nationalists had also moved toward the still newer ideals of socialism. Socialist nationalists criticized Mazzini's vision and leadership, and spoke even more explicitly of plebeians as part of the Italian nation. Melding Mazzini's faith in insurrection with Saint-Simonian dreams of economic equality, Carlo Pisacane, Giuseppe Garibaldi, and lesser-known men like Francesco Vigano and Pasquale Salvi linked national independence with solutions to the problems of Italy's poor, notably land reform.[25] They called for political changes to eliminate the gaping chasm between Italy's two races. Before 1860, their numbers remained small. In exile, however, the influence of radicals was considerably greater. Diaspora nationalism would not be the mirror image of the movement in Italy, in part because of the character of the exiles, and in part because of the labor migrants they encountered while abroad.

One diaspora or two?

An important and sizeable group, exiles were one of several growing migration streams leaving Italy in the early nineteenth century.[26] The differing motives of exiles, labor migrants, and Italy's traditional migrants, the culture-bearing elite, scattered the three groups in differing directions. Of the three, furthermore, only the exiles saw labor migrants as part of the Italian nation or as participants in a national diaspora. Their efforts to "make Italians" abroad are notable mainly because they preceded comparable efforts in Italy.

Few observers of the nineteenth century attributed the restlessness of Italy's common people to the political ferment of the Risorgimento, and most states in Italy did not even count their border crossings. Nevertheless, the years of the Risorgimento seemed ones of fundamental change in centuries-old patterns of local and regional migrations.[27] The representation of plebeians in Italian long-distance migrations increased, as did the total numbers of migrants leaving Italy. Chimneysweeps and organ grinders appeared on the streets of Paris and London, and the Americas were already a popular destination for the poor by 1830. Records from the U.S. tell of the arrival of about 1000 "Italians" yearly by the late 1850s; Argentina's

Table 2.2 Destinations of Italy's migrants, 1789–1871 (%)

	*Elite**	*Exiles*†	*Labor migrants*‡
Europe	78	90	28
Africa	6	3	11
Asia	4	3	4
North America	4	1	9
South America	7	3	47
Australia	1	–	1
(Number	677	2839	549,465)

Sources:

* Ugo E. Imperatori, *Dizionario di italiani all'estero (dal secolo XIII sino ad oggi)* (Genoa: L'Emigrante, 1956).

† My estimates from Rosi, *Dizionario del Risorgimento*, and Ercole, *Il Risorgimento italiano.* The figures are the emigrations of 1940 exiles, most of whom fled abroad on more than one occasion.

‡ Leone Carpi, *Delle colonie e dell'emigrazione d'italiani all'estero, Sotto l'aspetto dell'industria, commercio ed agricoltura* (Milan: Lombarda, già D. Salvi, 1874), vols. 1–2.

gate-keepers noted three times that number arriving yearly in 1857–60. Numbers emigrating to both countries rose sharply after national unification in 1861. In 1870, 23,000 Italians arrived in Argentina, and almost 3000 in the U.S.[28] When Italian statisticians began making their own estimates in the 1860s, about 100,000 emigrants were leaving the country each year. Assuming high rates of return, and migrations of even half this volume, as few as one million or as many as two million Italians left the peninsula and Sicily between the French Revolution and 1876 (when official Italian record-keeping began). In 1871, nationalist and former exile Leone Carpi attempted the first detailed census of Italians outside the country. Of a population of roughly twenty-seven million, half a million lived abroad (see Table 2.2). The largest group of migrants by far lived in South America, mainly in the Plata River cities of Argentina and Uruguay.[29]

Italy's traditional migrants, the elite producers of civiltà italiana, also continued to leave home, but they migrated along traditional paths. Performers of the "high arts" (as opposed to street musicians) remained the largest group (22 percent), followed by scholars (17 percent), literati (16 percent), and technical specialists — doctors, engineers, scientists (12 percent). As Table 2.2 reveals, most (78 percent) went to Europe, as they had in the past. In the early years of the century, only 2 percent of elite migrants went to the Americas; by the 1870s, however, the proportion had climbed to 20 percent.[30]

In their choice of refuge, as Table 2.2 shows, Italy's nationalist exiles stayed even closer to home, undoubtedly so they could easily return to continue their political struggles for liberty or independence. Three-quarters of over 1900 exiles returned to Italy, usually after relatively short absences. Unlike their elite and labor migrant counterparts, their migrations were staccato bursts accompanying the political crises of the Risorgimento: large groups left Italy in 1821, 1830–3, and in the aftermath of the widespread revolts of 1848–9. Exiles preferred France (25–35 percent of those going abroad), Switzerland (about 10 percent), and the countries around the Mediterranean basin (10–20 percent). Despite the liberal and republican revolutions that had occurred in the New World, the Americas never attracted as many as 10 percent of Risorgimento exiles.

Elite and plebeian migrants of this era traveled through differing social networks to differing destinations. Even when they lived in the same places, they were parts of separate diasporas. Carpi's survey found settlements sharply divided between a core group of settled business, professional, and artisanal workers (the urban elite that sometimes called themselves prominenti or "notables") and a transient population of laborers and men the consuls termed adventurers. Where they settled among labor migrants, nationalist exiles did, however, set to work to create united Italian settlements ("colonies," Carpi called them) of rich and poor. In those few places where they mingled with the labor migrants, they became what Emilio Franzina has recently termed living agents of Italian nationalism.[31]

Two factors pushed exiles toward a diaspora nationalism rooted in cross-class alliances: their political ideology and the ways they supported themselves financially. Numerically, many more democratic and revolutionary than moderate nationalists took the path of exile. Of over 1900 exiles, the largest group had participated in revolutionary actions (707), most of whom (681) called themselves republicans. (Only twenty-two declared themselves socialists, and 148 limited their activities to support of Cavour's moderate policies.) These were men radical enough to have used or supported the use of violence in pursuit of political change at home. Many were also followers of Mazzini. For them, the movement for Italian unification was a way to pursue an international program of republican insurrection that they imagined would topple autocrats and ancien régimes everywhere.

The pragmatic search for paying clients also drove many exiled doctors, teachers, lawyers, and pharmacists toward the labor migrants. Businessmen exiles were deeply involved with the labor migrants of their era. This was especially true of the merchants who imported, exported, or organized shipping and trade: these were the employers, transporters, provisioners, and ticket agents for the labor migrants. Over thirty exiles of the Risorgimento — men like Pietro Bosso and Bartolo Ghesa — supported themselves abroad as engineers and sub-contractor builders of railroads or canals. Years later, such padroni (bosses) gained notoriety by recruiting thousands of Italian laborers for similar work.[32]

The migrations of the Risorgimento thus had a somewhat complex effect on relations between the prosperous and urban nationalists and the poor and rural majority of Italy. Migration separated the Europe-bound exiles from the far more numerous mass of America-bound labor migrants, geographically perpetuating the cultural chasm that already separated the two groups at home. At the same time, however, political persecution sent into exile those nationalists most likely to seek allies among men of humbler origin. Migration transformed exiles into potential leaders of diaspora communities because exiles themselves viewed labor migrants as fellow Italians and potential followers. Together, exiles and labor migrants did on occasion cooperate, and when they did, they created a diaspora nationalism that shaped, but ultimately could not determine, events in Italy.

Diaspora nationalism: Italians made abroad

It is quite possible that peasants and workers more often came to think of themselves as Italian while abroad than at home. Even in the Renaissance, foreigners had failed to differentiate among Italians' many local loyalties. Risorgimento activists' public and internationally recognized campaign for change at home encouraged natives of France, North Africa, and the Americas to view migrants in national terms, as Italians — so they were listed, for example, in the immigration statistics of the U.S. and Argentina. Already in the 1820s, migrants living in the Plata regions of South America sometimes eschewed regional labels and referred to themselves as Italians.[33] Perhaps this explains why plebeian migrants also more often became nationalists abroad than at home. In Paris, London, and Buenos Aires, small numbers of exiles and workers constructed two somewhat distinct varieties of diaspora nationalism. Each had its own impact on the creation of a new state in Italy.

The best-documented and most pacific forms of diaspora nationalism developed among exiles and workers in London, where Giuseppe

Mazzini lived most of his life as an exile after 1837. By the 1830s London had a significant population of ordinary labor migrants from central and northern Italy. In an Italian neighborhood off Clerkenwell Road lived ice cream and other food vendors, along with growing numbers of street musicians and street tradesmen, many of them children.[34] Until the 1820s, Londoners had known Italians — though exiles — as an elite, cultured people who included the circle of poets (Camillo Ugoni, Antonio Panizzi, Giovita Scalvini, and Carlo Pepoli) gathered around Ugo Foscolo. Even prominent military officers — like Carlo Beolchi, Luigi Gambini, Giuseppe Vimara, and Evasio Radice, who had planned the San Salvario constitutionalist revolt in Turin in 1821 and then fought in Spain before settling in London — found welcome there.[35] By comparison, respectable Londoners found the poverty of the new child workers horrifying.

Once Mazzini had secured a precarious living for himself as a writer and journalist in London, his nationalist sensitivity to negative stereotypes of the migrant laborers among the English pushed him to action. Mazzini possessed an abstract sympathy for plebeians' membership in the Italian nation and he wanted to defend the good name of all Italians by finding Italian solutions to the problems of Italian workers abroad. His first act (in 1840) was to form a self-help society among humble artisans in London as a branch of Young Italy — much like the Unione degli Operai Italiani (Union of Italian Workers) which had already formed as a branch in Paris. (An Italian cabinetmaker and bookstore owner in London, Pietro Rolandi, had drawn inspiration instead from the English Chartists to form a similar organization of artisans the year before.) In 1841 Mazzini opened a school where working men and child street musicians were taught Italian, arithmetic, geography, and, later, English and mechanics. Lectures focused on contemporary events in Italy and the movement for national unity. Mazzini later moved his house to be nearer the school in Hatton Garden.[36]

Mazzini's modest experiments in workers' education and self-help were London's contribution to diaspora nationalism; only in the 1850s and 1860s would such initiatives become popular in Italy. Nationalist workers' societies and self-education projects in the diaspora thus formed one important foundation for Italy's emerging labor movement.[37] Comparable diaspora initiatives also developed elsewhere, notably in the Americas. In Buenos Aires, exiles found support among the artisans and petty traders of that city. Argentina's first Italian-language newspaper was aptly titled *Nazione italiana* and republican exiles subsequently founded the first mutual benefit society for workers, Unione e Benevolenza, in 1858.[38] The use of "Italian" and "nation" in diaspora institutions was significant; national identity took precedence over the ideologies of their founders. Equally significant, however, were terms like "union," "mutuality," and "benevolence," which linked Italian nationalism to the economic concerns of ordinary migrants, and to their lack of interest in accepting charity from the Catholic church. The combination worked. About a third of the Mazzinian nationalists among exiles had not left Italy for political reasons at all; they were labor migrants who became republicans and nationalists while living and working abroad.

Mazzini's dream — of plebeian soldiers fighting in nationalist insurrections under republican leadership — became a second important form of diaspora nationalism. It developed not in London under his leadership, but rather in Paris and in South America. France, with its revolutionary past, exercised a fascinating attraction on Italy's nationalists. One in five exiles of the Risorgimento spent at least some time in Paris. The earliest included soldiers and political activists fleeing the republican revolts of the 1790s, the supporters of Napoleon's Italian monarchy, and — after the Restoration — the organizers of conspiracies and constitutionalist revolts. Already under the ancien régime, Paris had begun to draw workers from Savoy and mountainous northern Italy. By the 1820s labor migrants from North and central Italy included chimneysweeps, street musicians, traveling salesmen, magicians, animal exhibitors, and street vendors of glass and plaster figurines. After bread riots and revolution in 1830, Parisian police became alarmed at the arrival of hordes of still poorer Italians, including child street musicians, who crowded on the Left Bank, between the Pantheon and the Jardin des Plantes.[39]

Moderate exiles and labor migrants lived apart in Paris. The moderates typically found respectable employment with Italian banks or French universities or newspapers, or else they taught Italian to middle-class Parisians. A surprising number — men like Giulio Dragonetti and Enrico Montazio — worked with Parisians on joint publishing projects, for example the literary paper *Revue Franco-italienne* and its business counterpart, the *Courrier Franco-italien*. Exiled lawyers and government officials more often depended on patronage and support of prominent, wealthy, and sympathetic Italian aristocrats like the Lombard aristocrat Cristina Belgioso, whose salon became a center for exiles. Exiles raised money and argued their political differences in their own Italian-language newspapers like *L'esule* (founded by Giuseppe Andrea Cannonieri) and *La gazzetta italiana*.[40] This pattern of class segregation emerged in New York, too. There, a first "congrega" (or section) of Mazzini's Young Italy movement had formed already in 1841. Slightly later, Secchi de Casali, a native of Piacenza who had earlier fled police persecution to Europe, Africa, and Asia, founded the Italian-language newspaper, *L'eco d'Italia*. It faced competition from a radical republican and Mazzinian newspaper, *L'esule italiano*. Thoroughly bourgeois, New York's exile institutions were vigorous — they created a $10,000 war relief fund after Vittore Emanuele's 1859 war with Austria — but had little impact at home in Italy.[41]

When radicals arrived in Paris, their initiatives focused more often on insurrection, conspiracy, and assassination, forging a bridge to the labor migrants. Drawing on the tradition of revolutionary conspirator and Jacobin Filippo Buonarroti, Italian nationalists saw Paris as their most important center of revolutionary activity. Radical exiles had made Paris an important base from which to organize armed invasions of Italy already in the 1790s. In 1799 exiled soldier Giuseppe Rosarol had found humble migrants in Paris willing to enroll in his miners and plowsmen brigade and to return to Italy to defend the Cisalpine republic from Austrian invaders.[42] Somewhat later, the early socialist and long-time exile Giuseppe Ferrari wrote his "Philosophy of Revolution" while in Paris.[43] In the Parisian revolts of 1830, an Italian Revolutionary Committee supported the French in arms against Charles X, and Italians of diverse backgrounds fought on the Parisian barricades (as they would again in 1848 and 1871). In the 1820s and 1830s, Italian exiles Celeste Menotti and Felice Argenti organized revolts in Italy from their base in Paris. The central Italian revolts of the early 1830s, along with repeated later invasions aimed at sparking insurrection at home (and supported by Mazzini from London), typically began among exiled plotters in Paris. Later in the 1830s conspirators found supporters in the Parisian Union of Italian Workers started by Mazzini to attract working-class migrants to nationalism.[44]

In the 1840s and 1850s, exiles frustrated by Louis Philippe's and (after 1848) Louis Napoleon Bonaparte's repeated interference in the governance of Piedmont and the Papal States turned to humble migrants like Giovanni Pianori and Giuseppe Andrea Pieri as assassins. Pianori, a shoemaker who had fought in Rome in 1849 and murdered his wife's lover, was guillotined after an 1855 attempt on Napoleon's life. Pieri (a hat maker and friend of Felice Orsini) had already fought in the French Foreign Legion and been expelled from the Union of Italian Workers for undisciplined behavior. He was arrested with a bomb in 1858 and executed.[45] While provoking considerable controversy, the recruitment of plebeian assassins in Paris was not a unique development of the diaspora but had precedents in Italy.

Only in South America did the recruitment of ordinary laborers for nationalist insurrection arguably become a mass movement. There, the charismatic Giuseppe Garibaldi successfully attracted large numbers of ordinary workers as soldiers. Tellingly, Garibaldi's activism earned him the sobriquet Hero of Two Worlds, for his earliest recruits fought first for liberty in the Americas, and only later as Italian nationalists.[46] Still, among Italy's Risorgimento nationalists, only Garibaldi created a mass movement of plebeian followers — his "garibaldini." The garibaldini (535 men) were the third largest group of activists among 1900 nationalists who lived abroad; 15 percent were workers and peasants. Many met Garibaldi in exile; they were labor migrants who became Italian nationalists in South America.

In sharp contrast to Paris and London, the coastal areas of South America in the 1820s and 1830s were raw and unsettled. Following their recent escape from Spanish and Portuguese domination, their economic and political futures were open and violently contested. Genoese sailor merchants had long ago reached this corner of the Spanish Empire; Ligurians and Genoese continued to do extensive business here after the American revolts against Spain. Migrants from Italy had a small presence in the South American countryside, too, especially in the southernmost province of Brazil, Rio Grande do Sul.[47]

Garibaldi himself grew up as a Provençal-speaking child of a small-scale merchant-sailor family in Nice, a city then dominated by people originating in Italy's northwest. Becoming a sailor, Garibaldi arrived in South America after years of travels in the Mediterranean (where he had ventured as far to the east as Constantinople and the Black Sea). Garibaldi claimed to have learned of Saint-Simonian socialism, to have joined Giovine Italia, and to have adopted the republican ideas of Mazzini before leaving the Mediterranean. He arrived in Rio de Janeiro in 1835 and immediately followed the paths pioneered by earlier military exiles from Italy in Paris and Mexico. By 1836 he had organized troops of migrants to support the revolt of Rio Grande do Sul against what he called the Portuguese imperialism of Brazil. Garibaldi subsequently organized an Italian Legion (its model was the French Foreign Legion) in Montevideo.

They fought with the rebel colorados against the blancos (forces of the dictator General Juan Manuel de Rosas in Buenos Aires). In these early campaigns the garibaldini adopted their emblematic red shirts, and Garibaldi became famous as "el diavolo" — a daring guerrilla general whose Creole wife, Anita, followed him into battle.[48]

In Latin America, Garibaldi's battles made him a supporter of local autonomy and democratic republicanism. His enemies were the native Creole elites with their desires for a state they could more easily control once centralized in Argentine and Brazilian cities. Ready to fight under the universalist but revolutionary banner of "equality, liberty, and humanity," Garibaldi returned to Italy with a retinue of trained soldiers in 1848. He arrived in time to fight against the Austrians in the North, and to defend the Roman Republic from foreign invaders before again fleeing into exile in the Americas in 1849. Ironically, Garibaldi's European campaigns unified Italy under a single, central, dynastic monarchy — that of Piedmont — and its ruler, Vittorio Emanuele II, the King of Sardinia.

Typical of Garibaldi's early recruits in South America was Luigi Rossetti, born in Genoa in 1827. First a sailor, and later a shopkeeper in Rio de Janeiro, Rossetti fought (and died) with Garibaldi on his ship *Mazzini* in the Rio Grande do Sul revolt. Among plebeian migrants who then followed Garibaldi back to Italy to fight in 1848 was Giacomo Minuto, born in 1819 near Savona. Minuto joined Garibaldi's Legione Italiana in Montevideo, then died fighting with him in defense of the Roman Republic in 1849. A second recruit returning from Argentina was Andrea Sisco, formerly of Bastia, who fled Italy after the fall of the Roman Republic, and then worked again as a laborer in Montevideo. In Italy, too, Garibaldi attracted plebeian followers. Typical was Antonio Elia, born in Ancona in 1803. A sailor and laborer, Elia participated in the Ancona revolt of 1831. In 1848 he organized a strike of sailors, and then joined Garibaldi's subsequent campaigns — for which he was later shot as a traitor.[49]

Mainly illiterate, the garibaldini and their political aspirations are difficult to assess. Garibaldi himself was a charismatic leader and a man of action, not an ideologue of great coherence or depth; his moderate opponents portrayed him as uneducated, crude, violent, and as dangerous as the masses he sought to lead. His military campaigns and public statements reveal him as a republican and revolutionary but also as a nationalist willing to compromise with monarchy. As a leader of a mass movement, he sometimes seemed a populist, and he was not (as was Mazzini) lacking in sympathy for the poorest Italians' hatred of the rich. But his leadership was not democratic, despite his sometime support for universal manhood suffrage. It originated in military discipline and individual authority. Had Garibaldi made his reputation only in the Americas, historians might now label him a caudillo (or local strongman) who lost out to the centralizing elites of Argentina and Brazil.[50]

In Italy, the influence of Garibaldi on the course of the Risorgimento and the formation of a nation of Italians was complex. When Garibaldi returned in 1859, he and his Cacciatori degli Alpi (Alpine Hunters) first fought for the immediate political goals of the moderate, and monarchist, nationalists who sought unity under Vittorio Emanuele's leadership. His subsequent Mille campaign (which took its name from his 1000 soldiers) invaded Sicily during the anti-Bourbon revolts of 1860, and Garibaldi's promise of land to Sicilian peasants seemed an open invitation to humbler men to join the nation. The peasants' revolt insured Garibaldi's success but, thus empowered, he then offered southern Italy to the King of Piedmont, Vittorio Emanuele, whose general, Nino Bixio, brutally restored order without initiating land reform. The rebellious Sicilians had freed themselves from the Bourbons only to experience subordination to a new, equally foreign, monarch.[51] In the years after unification, Garibaldi at times submitted to Vittorio Emanuele, and to the Italian monarchy he had helped to create, but on occasion, as at Aspromonte, he also defied his king. His last campaigns, by contrast, again made him the defender of republican revolution, and a supporter of the French communards and the First International.[52]

Garibaldi's success in transforming labor migrants into supporters of nationalist activism changed the course of Italian unification but did not determine its outcome. The moderates and liberals who supported Cavour and Vittorio Emanuele's constitutional monarchy held firm control of the Italian state from the first. They quickly distanced themselves from Garibaldi's and Mazzini's more democratic expressions of Italian nationalism. Like Garibaldi, most Risorgimento exiles returned to lives of political marginality and disappointment.[53]

Garibaldi himself complained that Italy's rulers had discarded him and his followers "like an orange peel." In 1880, he spoke for many when he declared sadly: "it was a very different Italy which I spent my life dreaming of, not the impoverished and humiliated country which we now see ruled by the dregs of the nation."[54] In turn, Prime Minister and Sicilian Francesco Crispi — a nationalist activist, former aide to Garibaldi, and an exile in the 1850s — reprimanded his republican and socialist critics with a history lesson about the making of the Italian state. Crispi lectured peevishly that common people had to "remember that everything that happened in our country in this century was the work of the bourgeoisie; it is to them that we owe our unification."[55] Moderate nationalists had indeed created an Italy, and an Italian state. But they had failed to "make Italians" — that is, a nation to support that state. Without a unified nation, Italy's new state could not guarantee order or stability.[56]

Making Italians at home

For the first thirty years of Italian independence, plebeian discontent with the new state scarcely waned. The northerners dominating the new government saw rural disorder through a lens ground by their anti-clericalism, their enthusiasm for the civilizing mission of a secular civiltà italiana, and the new science of social Darwinism. Faced with Catholic opposition and rural discontent, they stigmatized peasants as criminal members of a racially inferior people who preferred the superstitions of religion to civiltà italiana. After almost twenty years of virtual civil war, Italy's national government appointed a commission, headed by the respected Count Stefano Jacini, to inquire into the origins of the country's rural problems.[57] In doing so, it documented both the irrelevance of the nation to Italy's peasants and the growing popularity of migration among them. During a new round of peasant revolts in Sicily in 1893 and 1894, the Italian government fell, and thousands of peasants organized in peasant leagues (called fasci) declared their allegiance to Italy's new Socialist Party (the P.S.I.). Martial law and police brutality — not political participation — remained the shaky foundations for Italian national unity. Emigration soared.[58]

The northern moderates who first dominated the new Italian state faced enormous challenges, only some of their own making. The country's population was poor and largely illiterate. The new state was already in serious debt from the wars that had created it. Few roads (and even fewer railroads) linked Italy's many regions; without these linkages, Italy's economy, too, was national only in name. Worse, in the years after unification, native industries collapsed in some regions, adding to local problems of underemployment and unrest.[59]

Disastrously, the new state had inevitably earned the enmity of both the Vatican and many ordinary Catholics. The leaders of the new Italy were fierce anti-clerics, determined to end the secular power of the Pope, and to make Rome (still under papal control in 1861) capital of a united Italy — a goal they achieved in 1870. One of the first acts of the new Italian government was to confiscate church lands, and to sell them to individual owners in order to raise money. In doing so, they eliminated the resources the church had drawn upon for charity work, and left thousands of clerics without work or income. Rule by the moderates posed bourgeois Catholics with difficult choices after Pope Pius IX (on whom some moderates had once pinned hopes for leadership of a united Italy) demanded that Catholics reject the new "godless state." Pius IX forbade them to vote — a prohibition that remained in force until 1905. The godless state meanwhile labeled as bastards any children of marriages sanctioned only by the church and not registered with the state. It was in hundreds of offices of the "stato civile" (civil registries), and on millions of birth, death, and marriage certificates, that the new state first tried to "make" Italians — many of whom saw the Catholic church, and not the new government, as the most important moral arbiter.[60]

During the 1860s and 1870s, popular opposition to Italy's new government focused on three other widely hated policies, instituted by the leaders who succeeded Cavour after his early death. The first was a new tax imposed on the milling of grains, the second an international policy of free trade, and the third the government's demand that every Italian boy serve in the country's army as an obligation of citizenship. The result was a veritable civil war in many of the southern provinces, where peasants attacked tax collectors and fled the draft into the mountains and caves of the interior. Even more fell into the hands of local security officials when they attempted to cultivate or to hunt and gather on lands once held by their commune or by Catholic religious orders, but now transformed into private property. In the cities, artisan producers found themselves impoverished when imported goods robbed them of their traditional customers. When they organized in pursuit of self-help or political influence, the government banned their associations as seditious. If they stole to support themselves or their children, they, too, landed in jail.

Italy's new leaders dismissed this civil war as mere criminality; they mounted a war against brigandage. Cavour's envoys in the South in 1860 had already written home in astonishment, "What barbarism! Some Italy! This is Africa: the Bedouin are the flower of civil virtue compared to these peasants."[61] Eager to join the nation, the Sicilian author of an influential book entitled *Contemporary Barbarian Italy* issued the most damning condemnation of all: the people of Naples, unlike other Italians, he wrote, were a "popolo donna" — a feminine people.[62] Such impressions held even though some government representatives recognized that poverty — not race or gender — was the fundamental problem in the countryside. Heading the campaign against brigandage in the South, General Pallavicini summed it up succinctly. "Delinquency," he wrote, "which in this province is synonymous with theft, is the war of the poor against the rich."[63]

Pallavicini recommended imprisonment or death for bandits and their relatives. By labeling the rebels brigands and thieves, however, Italy's rulers transformed the very men they wished to "Italianize" into racially inferior criminals.[64] Virtual martial law remained in effect in large parts of the South until the late 1860s. With over 200,000 Italians imprisoned yearly, Italy's first *Annuario statistico* sadly admitted in 1876 that this "shameful army" of criminals symbolized "a sad primacy which has never been equaled in this field by any other European nation."[65] Italy had jailed a higher proportion of its citizenry than any other European nation.

Faced with disorder and revolt, Italian moderates saw schooling and universal male military service as the best alternatives to prisons for building a new nation. Both met with opposition. Schools promised at least to turn illiterate speakers of dialect into readers and writers of the modern national language — the Tuscan dialect now relabeled Italian. Certainly, illiteracy in Italy was high, especially in the South, where few schools of any sort existed under Bourbon rule. In Sicily in 1871, 89 percent of residents of small towns could not read or write; ten years later the figure had scarcely changed.[66] (In Piedmont, by contrast, where the state had introduced mandatory elementary schooling in 1848, few lacked basic reading and writing skills by 1881.) Opposition to schooling was surprisingly staunch. Hard-pressed peasant parents were unwilling to dispense with children's labor. Local elites feared both the costs and the possibly revolutionary consequences of literacy, and they blatantly ignored national directives to open secular schools. Heads of municipal government complained that men who learned to read in the army became obstreperous and restless upon return, abandoning their "proper" submission to ruling elites.[67]

In 1876, Italy's moderates lost control of the national government first to a loose alliance ("the historic left" of republicans and other critics of northern constitutionalists and monarchists), and later to the parliamentary dictators Agostino Depretis and Francesco Crispi. In an important compromise that linked the interests of northern industrialists and some southern landowners, Italy jettisoned both free trade and the hated milling tax.[68] Attempting a new tactic, the new government undertook a scientific study of the sources of rural disorder, much as the earlier critics of the governmental policies of the 1860s had already undertaken as individuals.

The fifteen-volume report of Stefano Jacini and his commissioners contains the clearest possible evidence of the political, economic, and social weaknesses of a state that lacked the support of a nation. It revealed, first, the persistent shock and bewilderment of bourgeois Italians when faced with the realities of rural life. Commissioner Francesco Salaris reported from Sardinia, "I was in the position of a person who incautiously sets foot into a complicated edifice, full of corners, pitfalls, and twisting corridors — looking for, but not finding, an exit."[69]

Worse, in every part of Italy, investigators documented the distaste and contempt urban Italians felt for the peasants of the countryside. Wrote one, "to call someone a paisan (villano) or viton (montanino)" (all dialect terms for the peasants of Piedmont) was to call that person rough and uncivilized. In other areas, city dwellers addressed peasants in the familiar form otherwise reserved for children ("dare del tu") while peasants addressed them in return in the respectful and formal third person once reserved for aristocrats.[70]

Some of the writers for the commission clearly shared these urban prejudices; one reported from Lucania in the South that only "certain hysterical philanthropists" saw the miserably poor huts of peasants as proof of their exploitation.[71] Most preferred to see peasants as suffering, but stoic and resigned, good Christians, but too ignorant and childlike to solve their problems. Even the most generous commissioners did not believe peasants deserved a more democratic government. One observer wrote from Calatafimi (where Garibaldi had won a major battle with peasant support in 1860) that the agrarian class there "does not even know what the right to vote is, nor do they care to know."[72]

The findings of the Jacini commission left little room for doubt about peasants' hatred of the rich, yet the commissioners refused to hear this hatred, let alone imagine how to end it. Commissioners described peasants as diffident, inarticulate, or cautious in addressing them. When they directly quoted peasants, however, the poor spoke in straightforward and clear language. They blamed their misery directly on the rich and their governors who were, they said, "without exception, thieves."[73] The same commissioner who protested his inability to penetrate the mysteries of the countryside heard a Sardinian peasant ask him,

> To what have we been reduced? The servant of every master. We cultivate our lands, and plant them, but the fruits are not ours. We work, sweat, and suffer under the rains that drench us . . . under the sun that burns us, and what remains of our work, our suffering, our sweat? They leave us enough to nourish us only, and that badly. That's all.[74]

The members of the Jacini commission made concrete recommendations for governmental action that ranged from colonization of empty lands in North Africa to increased investment in roads and agricultural schools. Some commissioners called for a return to firmer family discipline and to the strict patriarchal authority said to characterize traditional sharecroppers in Tuscany. Most urged the government to appoint better teachers, to improve the moral education of peasants, and to view military service as an opportunity to teach peasants higher standards of cleanliness and order. Such advice differed little from the "white man's burden" that European imperialists offered to lift in their African or Asian colonies.

The commissioners also called on the Italian state to regulate — or as they tellingly put it — to "discipline" Italy's migrants. The conclusion to the commission's report praised the supposedly natural, traditional, seasonal, male-dominated, and temporary migrations to Europe (mainly from Italy's northern and central regions). It also warned of the disorganized migrations of ignorant southerners and Venetians (from the northeast) who left permanently, and in family groups, to settle across the Atlantic, draining the nation's strength.[75] While no commissioner was illiberal enough to recommend limits on a citizen's right to move about freely, the commission as a whole encouraged the Italian government to direct and to protect its migrating citizens.

This advice, along with most of the findings of the Jacini commission, fell on deaf ears. By the end of 1893, when revolt again blossomed first in rural Sicily, and then in the rural and urban North, emigration from Italy was quickly rising toward its historical peak, surpassing national rates of incarceration. Unable to vote or to find effective political solutions to economic problems at home, "Italians-in-the making" voted with their feet. In the years that followed, migrant Sicilians, Venetians, Lucchesi, and Calabrians had greater importance as workers of the world, and as builders of foreign economies, than as Italian supporters of their own new nation state. Italy's new bourgeois rulers had stigmatized a majority of the new nation's citizens as racially inferior, rebellious criminals; those stereotypes would long adhere to Italy's workers of the world. They helped guarantee that the mass international migrations of the next fifty years would generate not one Italian diaspora, but many.

3

WORKERS OF THE WORLD, 1870–1914

> 'na festa, seradi a l'osteria
> co un gran pugno batú sora la tola:
> "Porca Italia" i tastiema: "andemo via."
> (A party, in the tavern, with a hard fist he strikes the table-top: "Damn Italy," he curses: "let's get outta here.")[1]

It was an odd, and somewhat sad, coda to the Italian nationalists' passionate and long struggle for national unity. No sooner did Italy have its own state than Italians began to abandon it in record numbers. Italy's statisticians counted 16.6 million departures during fifty years, as the nation grew from twenty-seven to almost thirty-six million by 1921. Over one million applied to migrate between 1876 and 1885; over two million in the following decade; and over four million in 1896–1905. Almost six million applied for passports in the ten years before Italy entered World War I.[2] At the end of the war, another 2.6 million Italians declared their readiness to leave home. Not yet noting returners systematically, such statistics threatened a hemorrhage fatal to a fragile nation's health.[3]

To explain the rapid growth of Italian migration after unification requires a look as much at the wider world as at the troubled new nation of Italy. The lives of the rebellious poor in Italy had long been constrained and dissatisfying without sparking mass emigration. Nor was Italy the only weak or troubled new nation state confronting a rapidly growing population and a stagnant economy in the nineteenth century. Had there been no jobs for them abroad, few Italians would have ventured forth from their villages. But there were jobs. Explanations for Italian mass migration thus lie as much in global and local histories as in the national history of Italy.

A few European nations' search for power had already created a Europe-focused, global economy centered in the sixteenth-century Atlantic.[4] In this first Atlantic economy, imperial nations conquered native civilizations in the Americas and enslaved Africans to work there, while Italy (as Chapter 1 showed) declined to its status as a colonized backwater.[5] By the nineteenth century, British, French, and Dutch empires reached into India, much of Southeast Asia, and the Pacific islands. Italy in 1861 gained independence after centuries as a minor dependency of the Spanish, French, and Austrian empires, but it did so without much chance of changing its marginal place in the wider world.

Still, migrants from Italy achieved a kind of humble prominence in the global economy of the nineteenth century, and especially within its second Atlantic economy.[6] Three changes sparked mass migrations from Italy within this Atlantic economy. First, the emancipation of African-origin slaves in the Americas threatened the stability of empires dependent on extracting raw materials from colonial plantations and mines. At the same time, anti-imperial revolutions in the Americas created huge and new, but sparsely populated, national states that believed, along with the Argentine Juan Bautista Alberdi, that "to govern is to populate."[7] Defining their new nations racially, and usually in opposition to the indigenous peoples who had long fought against European expansion onto their territories, new states sought settlers and citizens, usually from Europe. Finally, industrial capital spread from its earlier concentration in the cities of northern Europe and Great Britain to those of the Americas, and to plantations and mines in newer colonies in Africa and Asia. This new migration of capital created millions of unskilled jobs around the globe.[8] Who would replace emancipated African workers, populate the plains of Argentina and the United States, mine the iron and coal demanded by new factories, build the canals, railroads, tunnels to transport them, and work in the factories themselves?

In an astonishing number of cases, the answer was the same. People born and living in economic insecurity in some far-away village would do almost all these tasks.[9] Only rarely — in a few densely populated parts of northern and western Europe — could the villages of nearby rural regions within a developing national economy supply sufficient numbers of workers. In northern Europe, migration as often meant regional urbanization as emigration abroad, as rural women and men flocked to cities to work as domestic servants and industrial operatives.[10] Elsewhere, employers in plantations, mines, railroads, and factories, as well as new states in possession of fertile lands, were happy that dissatisfied peasants in villages far away proved eager to travel farther afield in search of wages. Oddly, in what was also an age of intensive nation-building, centralizing state power, and nationalism, a burgeoning global economy respected few national boundaries. The market for labor in this global economy truly was worldwide.[11] Mobile capital had given birth to mobile labor.

The nineteenth and early twentieth centuries saw a veritable explosion of human restlessness. As many as 150 million human beings (of a total world population of about 1.5 billion in 1900) walked, or took trains or boats back and forth across changing political boundaries. They searched, Italians claimed, for "pane e lavoro" — for work and bread.[12] Work and bread, not a secure sense of belonging to a nation, were uppermost in the minds of most labor migrants. The nineteenth century quickly became not just the age of the nation state and of global capital, then, but of the "international proletariat" that Marx suggestively (and hopefully) labeled the "workers of the world."

By almost any measure, the migrations from Italy were impressive. The thirty million Indians leaving a large sub-continent between 1830 and 1930 outnumbered them but were a smaller component of the much larger Indian population.[13] Higher proportions of Irish than Italians left home permanently in the aftermath of the potato famine of the 1840s, but the absolute number of Irish migrants was much lower than either Indians or Italians.[14] Only the numerically much smaller migration of (approximately) nine million Chinese scattered as widely as Italians (although two-thirds of them migrated within Southeast Asia) or as often returned home.[15] Overall, no other people migrated in so many directions and in such impressive numbers — relatively and absolutely — as from Italy. And few showed such firm attachment to their home regions, or returned in such large proportions.[16]

Thus, while the rulers of newly independent Italy struggled to build a national economy and to make Italians, Italy exported people more successfully than any other product. Italy's countryside raised workers to build the agriculture, infrastructure, and industries of other countries. Rather than binding the Italian countryside more closely to Italy's ancient cities or modern agriculture or industry, Italy's migrants would be workers of the world. Unlike the workers Marx imagined, however, their lives unfolded as much within familiar social networks of kin, friends, and neighbors as in the factories that transformed peasants into wage-earning proletarians. Like past migrations from Italy, the proletarian mass migrations of the late nineteenth century created multiple diasporas that gave new meanings to the adjective "Italian" without initially aiding the Italian state in its efforts to "make Italians."

Finding work in a global labor market

The global labor market of the nineteenth century demanded workers with some, but humble, skills — frontier farmers, urban technicians, and craftsmen, and wage earners with muscle power. "Populating in order to govern," the new countries of the Americas generated migrations of peasant families unhappy with agricultural conditions at home; such migrations included men, women, and children moving and continuing to work together in family groups. Losing their supply of slave labor, plantation owners now demanded individual, and usually male, wage earners, while the urban middle classes of growing cities wanted female domestic servants and urban industry hired mainly men. A small number of highly skilled men planned the construction of factories, cities, and systems of transport while millions more built infrastructure and then operated machines as factory proletarians.

Two-thirds of the migrants who left Italy between 1870 and 1920 were men — and men with traditional, but still important, skills. Peasants were half of all migrants before 1896. After that date, the representation of braccianti (landless wage-earning laborers) increased significantly, and the proportions of artisan and industrial workers also increased slightly. Together, manual laborers constituted over 90 percent of Italy's migrants; their manual skills, however, were still very much in demand around the world. Most of the rest of Italy's migrants were petty merchants, often with artisanal backgrounds and skills.[17] While male sojourners predominated, significant streams of family groups also left Italy, although not all traveled to frontier farms, as we will see.

Italy's elite continued their centuries-long migrations in small numbers, too, but the character of their migrations changed in telling ways during the mass migrations. Performing artists remained the largest group while other occupational groups well represented among the Risorgimento exiles — literati, soldiers, and scholars — now stayed at home, content with their new nation and its state. Replacing them in the second and third ranks of elite migrants were merchants and businessmen (15 percent each) and explorers (10 percent, mainly traveling to Africa). The newest exiles from Italy were men of the religious orders (7 percent) who — stripped of property and traditional privileges by the Italian government —

sought work ministering to migrants and adherents of new revolutionary ideologies, mainly anarchists and socialists. Elite migrants included a substantial proportion of the upwardly mobile among them. Fully 17 percent of Italy's new migrant prominenti were children of workers, craftsmen, peasants, and laborers. The smallest group had gained professional training as doctors and lawyers before leaving Italy. Many more became successful entrepreneurs by employing labor migrants in foreign enterprises (notably in construction and food industries). Thus a larger proportion of Italy's elite migrants now had jobs that bound them to the labor migrants rather than to wealthy consumers in foreign lands.[18] Their growing importance reminds us that migration, too, had become a big business.

For a global labor market to function, humble and often illiterate peasants from small and isolated villages somehow had to find their way to jobs thousands of miles from home. How could they know where to go? Most turned to familiar faces for information, advice, and assistance. They depended on experienced, returned migrants to guide them. Information about work abroad flowed through several channels. One was formed by networks of men sharing the same trade; these had developed inconspicuously over the centuries as towns developed a "mestiere per partire" — a trade that facilitated seasonal migrations abroad.[19] Labor recruiters returned from abroad provided the second source of information, especially for men without specialized trades. Over time, kin, friends, and neighbors became the third channel of communication. They provided the same services as labor recruiters — and often replaced them as the most important facilitators of successive migrations of the unskilled.

Long-distance migrations of both artisans and peasants originated in traditional circuits with deep roots in the first and second Italian diasporas. Peasants and shepherds had long descended from mountainous homes in the Italian Alps to go to Savoy; the surrender of parts of Savoy to France instantaneously transformed regional into international migrants. In the South, peasants from Piedimonte d'Alife had long gone to the plains of Calabria as harvesters; in the 1870s, they first learned from a returner there about work opportunities in America.[20] At least since the 1700s, too, towns in northern and central Italy had trained chimneysweeps, masons, and street traders as "girovaghi," who traveled abroad in small groups of three or four men. Returning home annually, they spread news and information about opportunities for work and life abroad. The returner thus figured prominently in many migrants' recollections of their decision to go abroad to work. Fortune Gallo from Puglia, for example, remembered "a big man, he dressed the part of the successful tycoon. His clothes were of the finest material and across his massive chest dangled a heavy gold chain."[21] Knowledge of the wider world was capital, and those who returned from abroad used that capital to build small businesses as advisers and moneylenders.

Girovaghi, like earlier political exiles, became knowledgeable labor agents or "merchants of flesh" as demand for labor increased worldwide. If Italians could sell plaster figurines all over Europe, could they not also sell men? To a very real extent, labor agents and recruiters like Nicola Cassinelli or Giovanni Veltri — called padroni (literally patrons, but usually translated as "bosses") — constructed the global market for the least skilled of Italy's laborers.[22] Padroni knew where jobs were located, and they often loaned workers the cash they needed to travel there. Observers were quick to see parallels between such business and the slave trade of earlier centuries. Because the abolition of slavery proceeded slowly over the century — beginning in the British Empire in 1832 and ending in Latin America in 1895 — the differences between free and unfree labor remained unclear and unspecified through much of the nineteenth century. References to many kinds of workers as wage slaves abounded.[23] Marx himself referred to the "veiled slavery" of wage earners in Europe.[24] In this context, padroni appeared as the new villains of the global economy. As one worker complained from Massachusetts:

> He extracts five dollars from every unfortunate immigrant he can grab and put to work. . . . And that is not all. After two or three weeks he gets them fired so that new ones can be hired. The victims do not dare complain for fear of the worst kind of retaliations.[25]

In fact, most of the padrone slaves were neither slaves nor completely free. Exploited they often were, but they were never the personal property of the padrone businessmen. The migratory children whom contemporaries called "little slaves of the harp" (because so many worked as street musicians) represented the least free end of a long continuum between slave and free labor; most were legally indentured apprentices.[26] Apprenticeship had always given masters considerable power over their apprentice boys, and critics claimed this caused tears to be "shed by a thousand mothers who have been deprived of their children."[27] But it was the parents themselves who signed their children into apprenticeship, a common practice in all the trades of rural Italy. Girovaghi "leased" (indentured) children in their home villages to work as organ grinders, street traders, and street musicians. In this way, children learned a trade (a "mestiere per partire") that prepared them for a lifetime of seasonal migration, and parents received a small cash payment in exchange for losing their labor at home. Because the child migrants provoked such harsh negative criticism, the Italian state — generally hesitant to interfere with parental rights — quickly prohibited the recruitment of migrant apprentices in 1873. By 1890, the little slaves had almost disappeared. Children remained a small proportion (10–11 percent) of Italy's migrants.[28]

New criticisms of adult slaves replaced the outcry about indentured children, however, especially in Canada and the United States. Certainly padroni, including men like Canada's Antonio Cordasco — who was crowned king of Italian laborers in Montreal in 1904 in an elaborate public ceremony — found jobs for millions of migrants from Italy, especially in construction.[29] But few of the men they recruited were under formal contracts of indenture. Most padroni were merely labor agents who exacted a commission from their clients. They sold jobs to men eager to migrate. They had no contracts of indenture, no contract laborers, and no slaves to sell to employers.[30]

Still, migrants from Italy were contract laborers in another sense: many worked for contractors and drew their wage directly from these small businessmen. Contracting had a long history in the Italian countryside, and both Po Valley women rice workers and southern wheat harvesters worked under contractor capi or bosses who found them seasonal jobs.[31] Ancient circuits of migrations to the south of France also sent women and men working for contractors as harvesters of grapes, flowers, and saffron.[32] In North Italian towns, engineers and master masons took sub-contracts for segments of construction projects in Italy and Switzerland. They recruited, supervised, provisioned, and paid work crews — including masons, stonecutters, and manual laborers — from their home villages. Sub-contracting had characterized large construction projects already in the age of canal-building in the early nineteenth century.[33] Many padroni later in the century were sub-contractors building railroads, tunnels, and the Panama and Suez canals with crews of migrant laborers. Some were engineers; others had technical degrees; still others were master craftsmen. Because contractors advanced cash to their crews for travel and then deducted interest along with payments for food and housing, their laborers often received pitifully small cash wages. Photographs of gang laborers — shoeless or stripped to the waist posing beneath, or to the side of, their shirted, shoed, and hatted padrone supervisors — differ little whether taken in Austria, Switzerland, Brazil, Argentina, Toronto, West Virginia, or China.[34] Although poor and hard-working, Italy's padrone slaves had nevertheless chosen work in the contractor's gang. Contractors gave even unskilled hod-carriers or pick- and shovel-men a "mestiere per partire."

Other middlemen, too, financed the longest trips of labor migrants across the seas. In cash-poor rural Italy, moneylenders set up as bankers and as steamship agents. Francesco Ventresca — a typical man who considered migrating in 1890 — at first sought to borrow 26 dollars from a fellow villager named Paradiso. Paradiso offered to migrate with him and others, but expected to receive 100 percent interest after six months of work in America.[35] In Asia, where expensive, long journeys also forced poor workers to borrow for their passages, scholars wrote of a credit-ticket system of migration and distinguished it from coolie migrations of indentured laborers.[36] Many Italians in the 1880s and 1890s left with credit tickets, enmeshed in a form of debt peonage. However deplorable, debt peonage was all too familiar to Italian sharecroppers who often had needed yearly advances of cash and seed in order to work in agriculture.

Italian laborers were not slaves but rather "servi" (clients or dependants) to their padroni. Outsiders could not understand their subservience and the apparent willingness with which clients submitted to bad treatment at a patron's hand. In Italy, however, few saw patron–client relationships as limits on individual liberty. On the contrary, patronage was an integral dimension of Italian life at every level of society, including within the modern state itself. While patronage made Italy's government corrupt, it also linked the country's many geographical fragments to its new central government in Rome. Italians pitied as truly powerless those lacking patrons, not those possessing them.[37] In 1868, Italy's Minister of the Interior actually instructed mayors not to issue permission to emigrants to go to Algeria or America without a padrone.[38] He saw men working under padroni as protected, and as part of an orderly migration that would return men home, and strengthen rather than drain national resources.[39]

Receiving countries too sometimes viewed semi-free labor as positive. The relatively small numbers of Italian adults migrating under formal indenture responded mainly to state, not padrone, initiative. Beginning in 1820, but more effectively between 1880 and 1902, Brazil's government, for example, paid passages for hundreds of thousands of families (many from the Veneto in the northeast) who agreed to work off their debts over six years. By the 1890s, their destinations were coffee plantations (fazendas) in Rio Grande do Sul and Santa Catarina. Because coffee prices declined as cultivation expanded, many indentured families found themselves deep in debt even after their initial contracts expired, provoking the Italian government to prohibit further assisted migration in 1902. Many migrants then returned to Italy or moved on to urban jobs in São Paulo.[40] Less controversial in Italy (but provoking negative comment from Australians hostile to unfree labor) was a Queensland planters' experiment with indentured Italians. In 1890 planters hired padrone C.V. Fraire to recruit North Italian families to sharecrop for two years on their sugar plantations. The indenture of these families had a happier outcome. Most became landowning, if small-scale, producers of the crop in a region where the crop was relatively new.[41]

Concerns about padroni and their slaves peaked even as Italian middlemen lost their monopolies over knowledge of the wider world and over the cash needed to travel to jobs. Their slaves had risked indebtedness to accumulate cash but by migrating they also gained knowledge of a wider range of job opportunities abroad. They could escape their dependence on the padrone, and they began to travel as part of what scholars now call chain migrations.[42] In chain migration, friend followed friend, wife followed husband, or brother followed sister, often on a pre-paid ticket, with no dependence on a middleman padrone. By 1908–9, 98.7 percent of southern Italians entering the U.S. had a pre-paid ticket or the address of a friend or relative.[43] Only in construction, where contracting continued to organize an entire industry, did the padrone remain an important figure. And even in that occupation, family ties intervened. In 1890, Michele Altafini wrote from Rio Grande do Sul to his parents in Rovigo, asking them to tell a brother-in-law: "I desire the best for his family and that in Brazil, as a mason, he could earn much money"[44]

Chains involving family and friends evolved over several generations, as is shown by a careful study of one family from Valdengo (and the nearby textile town of Biella in Italy's northwest) by Samuel Baily and Franco Ramella.[45] The first migrant from the Sola family left Italy in the 1850s as a street trader. He worked at a variety of wandering trades in France, Argentina, Brazil, and Africa before he stopped writing home from Veracruz, Mexico, in 1883. The children of three of his brothers also subsequently migrated. In the 1880s and 1890s, the metal worker son of one went to Lyons, France, with some fellow villagers, while the peasant son of another went with padroni to Europe, Africa, and the U.S. The children of a third brother went to Argentina. The first two Sola brothers eventually returned to Italy, but the daughter remained in Argentina. The son of the metal worker (and grand-nephew of the first family migrant) then migrated to Cuba. The peasants' two sons (one of them a construction contractor) instead relocated to Buenos Aires to join their godfather. A daughter went with friends and cousins to New York, and then (after she married) to Paterson, New Jersey. The various branches of the family continued to write home to Valdengo and Biella. Lines of communication criss-crossed in the home village, facilitated by the ceaseless transience of their paesani (fellow villagers). This meant, for example, that the parents of one of the Buenos Aires Solas learned from a returner that their son had married before the letter announcing his marriage arrived at their door. Another recently arrived Sola was hailed on a Buenos Aires street by a man he did not know who turned out to have known his father — whom he resembled — in Valdengo. Several members of the Sola family returned, both permanently and for visits, and others arranged to have their bodies buried in their home village.

Already under padrone recruitment, villages like Valdengo had become the center of a social network resembling a diaspora, with a village "home" (paese) and workplace "satellites" (usually called "colonies") abroad (fuori). Chain migration changed only the character of the relationship between the village and its satellites.[46] It facilitated longer stays abroad, and more of these became permanent. Chain migration also encouraged the migration of women, suggesting that more migrants over time left as family groups. In the 1880s, women made up 15 percent and, after 1900, 20 percent of Italian migrants. In the years bridging World War I, Italian women made up 34 percent of migrants.[47] The migration of women allowed reproduction to take place in the diaspora, laying foundations for either diasporic identities or incorporation into the countries where migrants lived and worked. It did not create a single, interconnected or Italian nation unbound, however. Italy generated many proletarian diasporas, and for several generations most had their center in a single paese.

Paese, regione, and the global labor market

Proletarian diasporas centered in particular villages were different from those of the early nineteenth century when class more than region determined migrant destinations (see Table 3.1). Whereas the Italian migrations of the years before 1861 had created two largely separate diasporas of exiles in Europe and labor migrants in the Americas, elite migrants now traveled to roughly the same destinations as ordinary laborers. The single striking difference among migrants of differing class origins is that the carriers of civiltà italiana more often went to the less developed parts of the world, notably Asia and Africa, than did labor migrants.

The earlier diaspora of the exiles and the newer ones of workers of the world were connected, however. Given the predominance of migrants from northern and central Italy before 1876 (see Table I.1), it is not surprising that mass migrations of peasants and workers also began earliest in those regions. Table 3.2 shows that the inhabitants of a few northern regions (notably Piedmont and the Veneto) showed the greatest propensity to migrate in the aftermath of unification. After 1890, southern rates of migration caught up and then surpassed them. In the South, too, interest in migration varied, and people from Campania, Calabria, Basilicata, and Sicily migrated in far greater proportions than from Puglia or Sardinia. Scholars have generally argued that areas of fragmented land ownership, where peasant subsistence production was slowly giving way to a cash economy and commercial agriculture, produced the highest emigration rates.[48]

Table 3.1 Destinations of migrants by class, 1876–1914 (%)

	*Labor migrants**	*Elite*†
Asia	–	4
Africa	–	11
Europe	44	39
North America	31	25
South America	24	19
Australia	1	1
(Number	14,037,531	565)

Sources:

* Gianfausto Rosoli (ed.), *Un secolo di emigrazione italiana, 1876–1976* (Rome: Centro Studi Emigrazione, 1976).

† Ugo E. Imperatori, *Dizionario di italiani all'estero (dal secolo XIII sino ad oggi)* (Genoa: L'Emigrante, 1976).

Table 3.2 Annual rates of emigration by region, 1876–1914*

	1876–94	*1895–1914*
NORTH		
Piedmont	0.95	1.30
Lombardy	0.60	0.90
Veneto	2.60	3.30
Liguria	0.50	0.50
Emilia	0.30	1.00
CENTER		
Tuscany	0.50	1.00
Marches	0.10	1.60
Umbria	–	1.30
Lazio	–	0.60
Abruzzi-Molise	0.60	2.50
SOUTH		
Campania	0.60	1.90
Puglia	0.10	0.80
Basilicata	0.20	2.60
Calabria	0.70	2.30
Sicily	0.20	1.60
Sardinia	0.10	0.50

Source: Annuario statistico della emigrazione italiana (Rome, 1927), Table III.
Note
* Average number of migrants per year as percentage of total regional population in 1881 (for 1876–94) or 1911 (for 1895–1914).

Italy's lack of national unity, and its fragmentation into many regions and localities, may also explain why Italians scattered more widely than other European migrants. Between 1870 and 1914, four million Italians applied to migrate to the U.S. and Canada. Six million went to work in other European nations and three million traveled to Argentina and Brazil. The others ventured in small numbers to North and South Africa, and to Australia (see Table I.2). People living in different regions definitely chose differing destinations when they left home. Job seekers from the North were particularly likely to cross the Alps in search of work, while southerners more often looked around the Mediterranean or traveled to the Americas (see Table 3.3). Southern diasporas often linked several locations in the Americas to a hometown, while in northern diasporas connections to Europe and South America were more common.

Table 3.3 Destinations of migrants by region, 1876–1914 (%)

	Europe	*Americas**	*(U.S.)*	*Other*
NORTH				
Piedmont	63	36	(10)	1
Lombardy	66	33	(6)	1
Veneto	81	19	(3)	0
Liguria	19	78	(23)	2
Emilia	69	29	(12)	2
CENTER				
Tuscany	61	36	(17)	2
Marches	33	67	(20)	–
Umbria	69	30	(19)	1
Lazio	14	85	(71)	1
Abruzzi-Molise	13	86	(59)	1
SOUTH				
Campania	7	91	(64)	2
Puglia	15	81	(57)	4
Basilicata	5	94	(53)	1
Calabria	3	94	(48)	3
Sicily	3	92	(72)	4
Sardinia	29	40	(12)	31

Source: Annuario statistico, Table III.
Note: * Including the U.S.

When they looked at the wider world, Italians themselves saw migrations across the Alps to Europe (transmontane) and across the oceans (oltremare) as fundamentally different, but they rarely distinguished between individual countries within these two large regions. Uneducated Italians in particular saw all the world beyond the seas as one undifferentiated place — la Merica. They rarely reported exact locations when reporting their whereabouts to their municipal draft board, but noted only "in America." America, furthermore, was not necessarily the United States or even the western hemisphere. "La Merica" meant Mexico, Canada, Argentina, or Brazil. One woman who migrated to Melbourne before World War I even insisted, "I migrated to America. It did not occur to me that Australia was not in fact America."[49] The special lure of the U.S. — said to motivate many European migrations — scarcely surfaced in the world view of migrants from Italy.[50]

Face-to-face relationships also organized migration, and thus made village social relations a further determinant of migrant destinations.

Although most northerners preferred European destinations, the residents of tiny, coastal Liguria (the area around Genoa) continued their centuries-long attraction to the Americas; they were also more likely to go to Australia than other regional groups. (See Table 3.3.) Emigration from Italy's largest islands also revealed their insularity. While migrants from Sardinia headed for Africa, Sicilians (who lived even closer to Africa) preferred the Americas, and especially the U.S., where fully 72 percent migrated.

Padroni, fellow workers, and family all represented potential linkages between a village and the wider world, so a paese rarely sent all its residents in a single direction. The investigators for Jacini's agrarian commission were among the first to observe this pattern. From Tuscany, one reported that emigration had been a constant of the region's life: for over thirty years, Tuscans had emigrated to Buenos Aires, Montevideo, Maracaibo, Constantinople, Tunisia, and Algeria. Another commissioner described the southern town of Picinisco, already infamous for its export of child workers in the early nineteenth century. He reported that a third of the town's residents worked in Paris, another third in English cities, and the rest in Ireland, Scotland, Germany, Sweden, Denmark, and Russia.

Choices like these revealed deep cleavages of occupation even in relatively small towns.[51] Like the Sola cousins, peasants and artisans from the same town might migrate in different directions. From Belluno in the 1870s, for example, a Jacini investigator reported that agricultural day laborers and artisans emigrated temporarily to Germany and Austria, while small landowners went to the U.S. From Lombardy, stonecutters, masons, and figurine vendors traveled across the Alps and beyond, while peasants ventured to Brazil.[52] In the 1880s, artisans from Sambuca di Sicilia pioneered the way to Louisiana, Chicago, and New York. But over the next thirty years, peasants followed them only to the canefields of Louisiana and the construction sites of Chicago. After 1890, most artisans who left Sambuca instead worked as barbers and shoemakers in Brooklyn.[53]

Gender relations in Italy's many villages determined women's migration rates but not their destinations. Women migrated in significantly differing proportions, but these differences cannot easily be summarized as a difference between a more patriarchal South and a less patriarchal North. Table 3.4 reveals that women were much more likely to migrate from Sicily (where most men went to the U.S.) than from parts of the North like the Veneto, where women remained behind during men's seasonal migrations to Europe. We do not know whether the length of men's absences,

Table 3.4 Females among migrants by region, 1876–1925

Region	*% female*
Piedmont	22
Liguria	29
Lombardy	17
Veneto	16
Emilia	19
Tuscany	21
Marches	21
Umbria	18
Lazio	14
Abruzzi-Molise	19
Campania	27
Puglia	21
Basilicata	30
Calabria	19
Sicily	29
Sardinia	15

Source: Annuario statistico, Table V.

or some calculus of female work opportunities at home and abroad, determined such outcomes.[54] Impressionistic evidence suggests that northern females more often migrated as unmarried young workers in search of employment while their southern counterparts typically migrated as wives or children in family groups.[55] When Italian women left home, however, they traveled in exactly the same directions as the men of their families and hometowns. There would be no separate male and female diasporas like the distinct diasporas of Biellesi or Sambucari.

In Italy's proletarian diasporas, the most powerful magnets remained the home village, not the work sites of Europe or la Merica. Exactly how many Italian migrants returned, or how often they returned or remigrated, unfortunately cannot be known. Local studies suggest that the transalpine migrants from the North were most likely to return. Of men leaving the Veneto and Friuli in the 1890s more than 80 percent and 90 percent respectively returned.[56] But since the Italian government did not systematically count returners from the Americas until 1905, or from Europe until 1921, even tentative conclusions are impossible. The proportion of returners among long-distance American migrants — 49 percent between 1905 and 1920 — was impressive. After all, travel across the Atlantic still took eight days to three weeks, and it cost more than half the yearly salary of an average peasant.[57] Samuel Baily has estimated rates of return from 44 to 53 percent from Buenos Aires and New York around 1900.[58] Betty Boyd Caroli cites a contemporary study that found 54 percent of southerners, 50 percent of central Italians, and 40 percent of northerners returning from the United States.[59]

Whatever their exact numbers, the returners and the transients kept alive — and lively — the flows of information between a paese and its diaspora. After arriving in São Paulo (Brazil) in 1891, a man wrote his sister in Milan to "tell Niulla that I found his son in San Paolo and drank a beer in his company and with Giacomo Savezza from whom I got your letter." He then added, "give my regards to all the printers [of his hometown — Rovigo — where his parents remained] and to Capovilla, too." Letter writers and returners alike reported on job opportunities, wages, new neighbors from the home village and the growth of their families, and on celebrations of religious festivals of patron saints of the paese homeland.[60]

Italy's proletarian migrants formed social networks in which the strongest bonds united several foreign destinations to a homeland village. A diaspora was never a transplantation of a whole community to a single site abroad. Few of the Sambucari living in Rockford moved on to Metairie (in Louisiana) or to Brooklyn, where other paesani lived, but most stayed in touch with relatives in their paese, Sambuca di Sicilia, and many returned there. The links between a paese and its satellites were strong ones; links among satellites were relatively weak. Love of a place in Italy — the patria — remained the most important source of belonging for Italy's migrants.

Migration thus helped keep alive the localism Italian nationalists sought to overcome. Traveling across the Alps, northern Italians learned the customs, languages, and economies of Europe. Southerners deepened their differences from them by learning the customs, languages, and economies of the Americas. Friulani and Sicilians already spoke mutually incomprehensible dialects: the differences grew as Sicilians learned to talk of "bakhousa" (outhouses) and "bordanti" (boarders) and Friulani of "bier." Sicilian demands for regional autonomy increased during emigration, and some islanders fantasized that Sicily might become the forty-ninth of the United States.[61] Meanwhile, northern Italians increasingly disparaged southerners in the language of scientific racism and emphasized what they shared with Europeans north of the Alps.[62] Distinctive migration patterns intensified Italian regionalism even as the nation state claimed to work to eliminate its influence. Emigration remained an important threat to the unity of a new and fragile nation.

National patterns in a world economy

Despite the survival, and even strengthening, of regional identities, the word "Italian" — as a description of Italy's residents — acquired clear but new meaning to outsiders as the workers of the world traveled the globe. Because migrants from Italy sought work from people they already knew, their migrations created a number of distinctive and distinctively "Italian" occupational niches within the global labor market. If the adjective Italian had come to describe an artistic people during the Renaissance, and a rebellious people in the early nineteenth century, in the modern world it described the Italian nation as poor, often ignorant, but "hardworking people."[63] The work they did worldwide differed from that of other national groups of migrants, but it also differed from country to country as each assessed how welcome Italy's migrants would be within its nation.

Italian migrants did not take jobs randomly, and as a result we can actually speak of Italian economic niches within a global labor market. By far the most important Italian occupational niche worldwide was that of male construction worker. Italians provided the bulk of the labor for the construction of Alpine railroad tunnels (Simplon, Gotthard, etc.) and lines between 1870 and 1920. They built railroads in Austria, and throughout the Balkans, and they traveled on into Asia, where they became 10 percent of the workforce on the trans-Siberian railroad. Thereafter, the Italian government's *Bollettino dell'emigrazione* warned prospective migrants against accepting work with contractors claiming to offer jobs in northern China. Northern Italian contractors and crews built railroads in North and South Africa and Argentina, and southerners provided the bulk of track repair crews in the eastern U.S. in the last twenty years of the nineteenth century. Italians labored on both the Suez and the Panama canals and on most major public works and private construction projects in American cities. Migration to Canada soared with construction of the Canadian Pacific Railroad in the first decade of the twentieth century. On five continents, Italian men were earthmovers, masons, and hod carriers — veritable human steam shovels who built the transportation and urban infrastructures of modern capitalism.[64]

Italians also figured modestly worldwide as laborers in commercial and plantation agriculture. In Louisiana as well as in Australia and Brazil, Sicilians temporarily replaced African and Asian laborers as plantation workers in the years 1880–1910. (In Louisiana, unlike Australia or Brazil, most sugar plantation workers were Sicilian men who had left their families behind.)[65] Padroni also recruited golondrine ("swallows" or "birds of passage") for work as seasonal harvesters on the Argentine pampas.[66] Directed migrations to the commercial vineyards and flower fields of southern France had a long history and continued into the twentieth century but involved both men and women migrants.[67]

The entrepreneurial spirit of Italy's medieval merchants also lived on among Italy's wandering street traders or girovaghi. In the modern era, large migrant populations demanding their own foods and services in their own dialects created ample business opportunities for grocers, restaurateurs, and saloon keepers.[68] For miners, too — a smaller but visible Italian specialty on four continents — ties of occupation, friendship, and kinship created a worldwide niche. Miners migrated in very distinctive streams. The sulfur miners of Sicily went in large numbers to coal mines in Alabama or West Virginia, while miners from Trento and Piedmont pioneered the way to Illinois, to Lorraine in France, and to Germany's Lothringen iron ore mines.[69]

Migrating mainly within family and friendship circles, and with fewer jobs open to them, Italian women also had a visible, important global niche in the so-called light or female industries. In the 1880s and 1890s, Italian women found work in Swiss and French spinning and weaving mills; somewhat later, they worked in similar mills in southern Germany.[70] In the Americas, Italian women worked in very large numbers manufacturing garments, cigars, textiles, and shoes. After 1900, they formed the largest group of workers in these industries in many cities in the U.S. and Argentina. Italian children joined them, completing a rapid transition from little slaves to industrial workers in thirty years.[71] Never having depended extensively on padroni, Italian women and children made the transition to industrial work more quickly than Italian men did.[72] Marx had imagined proletarians as male operatives; in Italy's diasporas, they were more often women and children.

Regardless of background, then, Italians performed different tasks in the global economy than migrants of other national backgrounds.

The clearest example is Italy's women migrants, who seem to have made the transition to industrial work more quickly than other female migrants. Irish, Finnish, Scandinavian, and African-American domestic servants flocked to the kitchens of the urban northeastern U.S.; Italian women did not. Domestic service was certainly familiar to them: in Italy, peasant daughters and wives had worked in middle-class kitchens, although few relished the task. But Italian women — unlike Poles — did not migrate to Germany, Switzerland, or France to work as domestic servants. Only in Latin America did Italian women work as laundresses and domestic servants, and then usually for Italian employers.[73]

In many respects, Italians' place in the global labor market resembled that of migrants from the colonies or dismembered older empires of Asia, especially China. Labor contractors, emigration agents, and indenture more often characterized Asian than European migrations, as did the accompanying charges of slavery and unfree labor. Even today, some scholars label the coolie trade of indentured servants that provided plantations with laborers as a new system of slavery.[74] Several million Indians, smaller numbers of Chinese, Japanese, and Pacific islanders, and a still smaller number of Italians went under indenture to Australian and American plantations to replace emancipated slaves.[75] Indians most often worked in the less free (indentured) and Chinese in the more free (credit ticket) recruitment systems. Italians generally avoided indenture, but they were among the very few Europeans to do any plantation labor at all.

Dependence on credit tickets was only one obvious commonality in the migrations of Chinese and Italian laborers. Like Italians, Chinese men worked in gangs for contractors who found them construction jobs, provided them with housing and food, and deducted payments for debts and subsistence before turning over cash as wages. The two groups established comparable occupational niches (in different parts of the world) as male specialists in construction and mining, and — to a lesser degree — plantation labor. Italian padroni had their equivalents in the Chinese crimps; both insisted that men indebted to them work off their debts through indenture, credit tickets, and labor abroad. Both migrations linked large numbers of small traders and common laborers. Finally, most Chinese and Italian migrants were sojourner men who returned home. Italian rates of return surpassed European averages (of about one-third) but were lower than those to China (which may have reached 90 percent).[76]

Italy's migrants' place in the global economy differed noticeably from other Europeans' in one important respect. The United States was not their preferred destination. The fact that migrants from Italy concentrated in somewhat different jobs in the economies of the many countries where they sought work helps us to see that the migratory workers of the world were not interchangeable laborers, unmarked by their national origins. On the contrary, national variations remained surprisingly visible in a globalizing economy. To some extent, migrants' localism helped create them. But the attitudes of receiving countries were also important. Some, far more than others, were unenthusiastic about welcoming migrants from Italy into their nations.

Unlike many of the other Europeans who showed a strong preference for migrating to the U.S., Italians did not settle on the frontiers of North America, nor did men take factory jobs there in large numbers. For several decades, men's jobs in the U.S. were too marginal to allow permanent settlement. Working seasonally, birds of passage built or repaired railroad lines as far west as Kansas and southern Texas. (West of these points employers turned to seasonal gangs of first Chinese, and later Mexican, workers.)[77] Even more important were seasonal jobs constructing the cities of the northeast, especially New York, but also Philadelphia, Boston, Cleveland, and Chicago. Each spring and summer, Italians paved streets and laid track for street railways. They dug tunnels for railroads, sewers, and subways, and laid the track for urban rail systems; they constructed urban water systems, including the giant series of reservoirs and aqueducts that fed New York City. They built bridges, factories, tenements, department stores, and skyscrapers.[78] In Louisiana, plantation laborers in the 1880s and 1890s managed to patch together year-round employment only by remaining on the move: they worked seasonally as sugar cane harvesters and grinders and then as trackmen for Gulf Coast rail lines.

Full-time jobs in the U.S. were harder to find. First northern and later southern Italians found more permanent employment in the mines of Pennsylvania, West Virginia, Illinois, Alabama, Colorado, Nevada, and Utah.[79] Only in California did significant numbers (mainly from Italy's North) become farmers and settlers, working as truck gardeners and wine and fruit growers.[80] In San Francisco and Tampa, northerners and southerners also found more permanent work as fishermen. Artisans and petty merchants also developed niches as produce vendors, bootblacks, shoe repairmen, barbers, and waiters.[81]

In only a few places in the U.S. (e.g. Tampa's cigar industry) did Italian men recently arrived from Italy become a significant component of local factory work forces, as women and children more often did.[82] By the twentieth century, shoe factories and garment shops also provided employment for Italian men, especially in New York City and Chicago. Sons of immigrants worked in larger numbers in factories as semi-skilled machine operators.[83]

In Canada, too, albeit somewhat later and in much smaller numbers, Italian men worked at first almost exclusively in construction. Thousands arrived in Montreal yearly to man the work crews building the Canadian Pacific Railroad. Although they created the steel road enabling settlement of Canada's vast prairie provinces, few Italians settled there — as did Scots, Irish, and Germans from Germany and Russia. Italian settlements instead developed in Montreal and Toronto and around the docks of the Great Lakes.[84] The newest of the settler colonies of the British Empire, Australia offered a slightly different mix of opportunities. There, Italians worked on sugar plantations (in Queensland) and in mines in the north, while in a few coastal cities they pursued the usual mix of urban and street trades. Still, Australia remained a distant and very expensive journey for Italians before World War I and few went there before 1945.[85]

Italian men's employment remained marginal throughout the English-speaking world and in the British Empire. British plantation owners in Asia and in the Caribbean, and British mine managers in Africa, recruited workers from within their own colonies in India and from their commercial sphere in south China.[86] At the metropolitan center of the empire, in England too, Italian street traders and indentured children arriving after 1830 never gave way to male migrations to industry or the mines. The Irish poor from Britain's nearest colony took those jobs.[87] The small population of Italians in England instead contained many small businessmen in the ice cream and food trades. Achile Pompa, a padrone from the hometown of the child musicians (Picinisco), had helped large numbers of new migrants get started in that trade in the early 1890s.[88]

Italians enjoyed far superior job opportunities in Latin America and proportionately more Italians than other Europeans (especially from northern Europe) migrated there. In South America, large numbers of male Italian sojourners worked seasonally and temporarily as harvest laborers and as builders of railroads and cities. But far more often than in the U.S. or Canada, Italians also became independent farmers (albeit more often as sharecroppers or tenants than as owners).[89] They also more often became investors and businessmen in the new urban industries of Buenos Aires, Rosario, or São Paulo. New arrivals found South American cities at the same low level of industrialization as in Italy. Openings for technical, skilled, and clerical workers — like the Sola brothers from Valdengo — were abundant because native elites disdained business and commerce. Migrants from Italy transformed artisanal shops into small manufactories of garments, leather, textiles, and machine goods.[90] Although British capital dominated in Argentina's large meat-packing and textile plants, some of Brazil's and Argentina's largest industrialists and a majority of their urban small businessmen were of Italian background.[91] Samuel Baily describes these Italians dominating the "middle sectors" of Argentina's urban class hierarchy.[92]

Italian-origin businessmen in turn hired newly arrived Italian immigrants as wage earners. Quite unlike Italians in the U.S. and Canada, Italians in Argentina and Brazil became the single largest component of the relatively smaller industrial working classes of these countries.[93] The Italians in Buenos Aires, in particular, created what sociologists would later call an ethnic or enclave economy where both capital and labor emerged from a single immigrant community.[94] Italians' economic status in Latin American cities more closely resembled that of the German, Russian, and Polish Jews of New York's garment-making enclave economy than that of the Italian migrants of New York. For Italian migrants, Argentina was more a land of opportunity than the U.S., and scholars have even called it "Italy's Australia."[95] In South America, Italians became modernizing forces in a way they could not in more developed economies.

The economic integration of Italians in more developed European nations fell between the patterns of Latin America and English-speaking countries. In Europe, as in the U.S., the largest Italian occupational group, especially in Switzerland, was in construction, with its sojourners in marginal jobs at isolated worksites under the control of contractors and bosses.[96] By 1900, however, the second largest group of Italian migrants in France,[97] and (slightly later) Germany[98] and Switzerland, worked in mining and industry. While they were a growing and increasingly important component of Europe's industrial working classes, migrant men also faced obstacles unknown in Latin America, for Europe's native workers already occupied the highly skilled jobs, blocking migrant mobility.

Marx's brilliant identification of the workers of the world notwithstanding, national differences clearly mattered in a capitalist world economy. They structured its market for laborers in significant ways. The employers of migrant labor were not indifferent to the national background or the gender of their workers. Migrant men and women from Italy thus filled a somewhat different range of jobs from either Chinese or German migrants worldwide, and they worked — and ultimately settled — in different countries, too.

The magnetism of the paese and the difficulty migrant men encountered in finding permanent work in many countries encouraged the persistence of high rates of return and repeat migration. Italy's diasporas were not to be short-lived social networks; they shaped the lives of many Italian families for several generations. Transnationalism thus became a very familiar, even ordinary, way of life for Italy's peasants and workers in the years between 1870 and 1930. Male occupational niches outside industry and transnational cultures that linked men and women within Italy's many diasporas explain why Italian migrants so often departed from the model behavior Marx predicted for workers of the world.

4

TRANSNATIONALISM AS A WAY OF WORKING-CLASS LIFE

> Cci dissi lu gadduzzu a la puddastra: Tuttu lu munnu è comu casa nostra.
> (Says the woodcock to the hen: the entire world is like our family.)[1]

In the years between 1870 and 1940, three generations of Italy's poor saw their lives transformed by repeated experiences of migration, life abroad, and return. Migration was a normal dimension of everyday life for the poor wherever they lived; it was no more of a crisis, nor was it accompanied by more intense discomforts, than any of the many hardships that poor peasants and laborers faced. In the 1890s diary of Andrea Gagliardo, a peasant who lived in the small village of San Colombano Certenoli a few kilometers from Chiavari in Liguria (Italy's northwest), references to America appeared almost daily. Members of Gagliardo's family emigrated for four generations — to Spain and Gibraltar between 1798 and 1825, to Argentina and the U.S. between 1870 and 1900, and to the U.S. between 1905 and 1925. In 1890, Gagliardo had relatives in California, Nevada, and Argentina, although he corresponded mainly with those in North America. He regularly received newspapers, news, and money from "la Merica" and he noted the births, deaths, and return of the migrants in his chronicle. America was an intellectual, financial, and social resource in the family life of Andrea Gagliardo until after World War II.[2]

As Gagliardo's diary reveals, proletarian migrants developed their own kind of cosmopolitanism during the mass migrations. Their transnationalism was not the high culture of civiltà italiana. It exactly resembled neither Mazzini's or Garibaldi's diaspora nationalism nor the internationalist aspirations of Mazzini's Young Europe or Garibaldi's commitment to republican liberation through armed struggle in Argentina, Italy, and France. Each village in Italy generated its own diaspora and tugged its migrants back home again. The tug was more powerful than patriotism for its roots were in the ties of family, sexual desire, and sentiment: proletarian cosmopolitanism was the product of each family's search for security in a global labor market.[3] Family security meant many things, of course — money, a dowry, house or land, a trade or profession, the comforts of a familiar dialect, cuisine, or religious rite, and the respect of friends and relatives in a face-to-face community. The reproduction of family life and the search for family security also established the limits to proletarian cosmopolitanism. Their family- and village-based cosmopolitanism made Italy's migrants quite unlike Marx's workers of the world.

To Marxist eyes, sojourners offer extreme portraits of proletarianization — men ripped from pre-capitalist communities, from the "idiocy" of rural life, and from the petty competitions of peasants Marx had portrayed as mere "potatoes in a sack," incapable of class-conscious solidarity. Workers of the world went anywhere to earn a wage, reduced to mere factors of production. They recognized no national loyalties, but confronted the terrifying power of capital in its harshest guise. Marxists expected proletarians would find solidarity among fellow factory workers and spark a working-class global revolution.[4]

In fact, of course, most of Italy's migrants remained economically and emotionally enmeshed in subsistence production and their rural communities. Most of the men were not factory workers, and national, cultural, and religious differences often divided them from their fellow workers. Migrants' lives responded as much to the reproduction of families at home as to the exigencies of wage-earning abroad. Working in different locations and in rather diverse jobs, Italian men and women occupied different class positions within the global economy and their everyday experiences of class also diverged sharply. Still, family life intimately entwined men and women, whatever their differing experiences. Wage-earning and subsistence production also proved quite compatible. Thus we should neither exaggerate the cosmopolitanism of the migratory men nor ignore the impact of the wider world on women in Italy's villages.[5]

For most families touched by migration, a transnational way of life meant mainly the construction of family economies across national borders.[6] Initially, family economies linked work camps populated by wage-earning Italian men (the "men without women") and rural Italian villages housing disproportionate numbers of women and children awaiting their return. Between 1870 and 1914, male work camps and rural Italian villages had more communication with each other than with the national societies that surrounded either. Because migrant families generally rejected permanent separations, or ones of indeterminate length, they faced one key question — whether to locate the family's home base in Italy or abroad.

After 1900, Italy's diasporas increasingly included what Italian speakers of the time called "colonies" to distinguish them from work camps. These were satellites — family groups and neighbors — of particular villages and regions. The homeland paese too transformed into a village of return as other families reunited there. Families experienced dramatic changes whether they relocated to Buenos Aires, New York, or Marseilles or reunited in Italy. While migrants abroad recreated elements of their older lives, returners seemed reluctant to respect many of the old rural mores. Transnational life began to create significant differences between the families and villages of Italy and the migrants transplanting themselves to new homes in the foreign "colonies" of their home villages.

Men without women

Canadian historian Robert Harney captured the peculiar quality of male proletarian cosmopolitanism when he called attention to Italian sojourners as the "men without women."[7] Living apart from women was no new experience for Italian men, although the distances traveled and the length of their separations increased with the mass migrations. Few saw sojourning as a desirable way to live but most accepted it as a temporary sacrifice to facilitate family security. And some found it adventurous.

Commentators in Italy like Luigi Villari used military metaphors to describe male migrations as "an army without generals."[8] Some men left begrudgingly, as they did also in response to the draft. The migratory men, too, called their sojourns "campaigns" from which they returned either victorious or defeated.[9] Observers labeled returning migrants "reduci" (veterans).[10] Like the draft, migration fulfilled hard obligations to the family and carried with it at least the possibility of adventure. Unmarried men particularly dreamed of "making America," and they saw their migratory campaign abroad as an effort to "far fortuna" or seek their destiny.[11] After 1861, military service to the nation and migration abroad quickly became new male rites of passage to adulthood, and ones not directly related to village and parental concerns about marriage matches, savings, or inheritance. By releasing sons from paternal supervision, both military service and migration allowed young men to make decisions as individuals and to earn individual wages. Both thus carried an implied threat to a father's power as cash became an increasingly important addition to peasant family economies. (The Italian state had forced peasants into the cash nexus with its demands for taxes, as did the spread of commercial agriculture.)

The search for cash to pay taxes or to purchase cloth, houses, or land gradually transformed peasant life throughout Italy in the nineteenth century. Peasants and artisans alike expected their children to begin working at the age of seven or eight; work was children's primary education. Married women and men both expected to work to achieve family security.[12] Even bourgeois Italians in 1871 did not always assume that their own family values — which made work an adult male obligation and excluded women and children from physical labor — were appropriate for the poor. In fact, middle-class observers idealized rural families that worked collectively under patriarchal authority as did the sharecroppers of central Italy, especially Tuscany. There a single head of household, usually the father (the capoccia), organized and supervised all family labor, with the assistance of his wife (the massaia).[13] Sharecropping households might include twenty persons who formed stem families (parents plus one married son and his family). Other households were joint ones where several married brothers cooperated under the leadership of their father or the eldest brother.[14]

As work for wages in commercial agriculture or in local industries became more common after unification, bourgeois observers wrung their hands over the collapse of traditional patriarchy, and with it the morality of the poorest. Landless families became more common, especially in the South where employment opportunities for women especially collapsed in the face of increasing competition. In northeastern Italy and in the Po Valley, a large rural proletariat existed by 1900, albeit of waged workers still employed in family groups.[15] In the northwest, wives and the youngest children farmed bits of land while older sons, daughters, and husbands sought cash wages. The Jacini commission in the 1870s blamed the draft and independent wage-earning for declining religiosity and deference among sons in the northwest; in the northeast they instead reported landless laborers living like animals. In the South, where agricultural families lived in large towns, far from their fields, men often traveled long distances to work while women worked on plots of land closer by, or earned wages through cottage industry.[16] In the South, men separated themselves from their families and from the institutions of urban southern life — church, home, shops, and piazza — for long periods in order to seek work on distant estates. A reporter for the Jacini investigation thus proclaimed women in the Abruzzi the real rulers of their families, a development that shocked them. Even a sympathetic Sicilian folklorist agreed, seeing matriarchal elements in island families, and in women's control of both their own dowries and a family's savings.[17]

Where men worked far away, the specters of female infidelity and male barbarism soon hung heavy over peasants desiring social respectability. Most peoples of the Mediterranean viewed female sexuality as a powerful and potentially disorderly force.[18] Men's limited control of female sexuality especially threatened family honor in the South because men in that region so often left their families to find work. Bourgeois and urban notions of civilized behavior symbolically reduced male peasants to the level of beasts that lived in the uncivilized countryside. Shepherds who worked long weeks alone with their animals seemed particularly dehumanized. In this context, male migration across the seas and mountains was familiar but it also raised familiarly troubling moral issues.

No one — Italian or foreign, young or old — found much to like about the all-male work camps where so many of the Italian hirondelles, golondrini, or birds of passage lived when they migrated abroad. For harvesters in Argentina or Louisiana, as for railroad crews anywhere in the world — but especially in Switzerland's mountains, and the emptiness of the North American countryside — work camps were far from urban centers. Italian workers in the camps felt themselves "fuori" (beyond civilization). They became "forestieri" (outsiders), not just as foreigners but as workers isolated from urban civilization. Hard physical labor for ten or more hours left men with little leisure and little entertainment. A simple meal and a few short hours separated labor from a night of rest in a bunkhouse or a dirty, crowded, railway car. On Sundays, men found no churches, families, or commercial entertainment within reach.[19] Thus, Italian sojourners might have agreed with the American who observed,

> Camp life is an unnatural life, and in it the coarse, vulgar elements of human nature come to the surface; the indecent story, the vulgar joke and the immoral picture are

> introduced and passed around. If intoxicants are within reach, the men will drink and gamble.[20]

Sojourners preferred work in cities, constructing subways, sewers, streets, and skyscrapers. Cane harvesters in Louisiana gravitated as quickly as possible toward New Orleans; harvesters left Argentina's pampas for Buenos Aires. There they found businesses offering to sell them the goods and services men without women might want. The café, tavern, or saloon, along with the boarding house and brothel — all often owned by the businessmen who worked as padrone labor recruiters, bosses and sub-contractors, and bankers — provided a sense of urban, if not familial, normalcy. Churches were few, but migrant men did not often claim to miss them. They instead reported excitement about the time they could spend in Montreal (which they preferred to staid, English-speaking Toronto) where excitement and entertainment could be had in the company of friends.[21]

Observers lamented the rude drinking, lounging, and gambling of these all-male communities. Oddly, however, few worried that "unnatural" (that is, homosexual) relations might develop among Italian sojourners, as observers did when confronted with Asian indentured servants on plantations around the world.[22] Sojourners themselves rarely complained of poor food, long hours, and communal sleeping. What they hated was their separation from women. Reported one, "no women in America, ostia — las' time I were six year in America, work in backwood, for two year never see a woman."[23] For a worker in Germany's brickyards the life of men separated from women meant a season of "voluntary imprisonment."[24] A great deal of anxious joking developed over improper relations between male boarders and the women (usually wives and daughters of padroni) who cooked for them. Regularly reviled by the native-born ("even a simpleton can see that they do not like us," one man noted[25]), the men without women surely patronized brothels. But only rarely did they find there prostitutes from Italy (except in Marseilles where observers reported women selling themselves in taverns).[26] Still, women in Italy rightly sang worried songs, "My love, my love, how far you are; Who's making your bed tonight? Who does, is not doing it well."[27] Unfortunately, few migrant men left evidence of their sexual activities, let alone thoughts about their significance.

Their silence probably reflected their fear that gossip about their behavior might have repercussions back home. Since most men both worked and suffered unemployment together with fellow villagers, they were never truly far from the social surveillance of the paese. It is unlikely that Italy's peasants expected sexual purity, or even fidelity, of men without women. Still, parents and wives alike feared losing the wages and loyalties of sojourners. Parents wanted their sons' marriages to remain under their control, at home. Transatlantic gossip worked well enough for a literate sojourner to write to his father pleading with him, "I would like to know who has been gossiping about me. Do not believe them. It is all lies."[28] (Significantly, he did not note what untoward behavior of his had sparked the gossip.) Robert Harney, too, emphasized the power of "the backbiting paisano or solicitous relative" over the man living apart from women. He concluded, however, that "filial piety, parental duty, and morality acted in the migrant to serve as a brake on his decline into brutishness." Even if the sojourner paid for sexual services, "He still worked to send money home for specific purposes."[29]

Rather than resent family obligations, sojourners interpreted their migrations as a fulfillment of those obligations. They viewed work itself — not their parents or wives — as the dictator of their lives. Claimed one, "My job was my via crucia, my misery, my hatred, and yet I lived in continuous fear of losing the bloody thing. THE JOB that damnable affair, THE JOB . . . this blood-sucking thing."[30]

Women who wait

The same hard work and community surveillance characterized the lives of wives and daughters, the "women who wait," as Caroline Brettell called them.[31] Whereas gossip about men may have focused on their care in earning and saving wages more than on their sexual propriety, the reverse was true of women. Their contemporaries called the women left behind "white widows," differentiating them from the black-clad widows whose husbands were not only departed but dead. White widows seemed more threatening to public morality than black ones because so many were young and of child-bearing age.[32] Statisticians in Italy noted the large numbers of married women with absent husbands, and moralists worried over the "young and beautiful bride who finds herself, only shortly after the wedding, alone, and almost a widow in her marriage bed. It is easy to imagine," this observer continued "what happens — and must happen — when 50,000 wives are condemned to forced retirement, while still full of the exuberance of youth, and often also very poor." The same observer claimed that men refused to consummate their marriages, hoping a virgin would resist temptation more easily.[33]

Despite predictions of moral anarchy, rates of illegitimate birth in Italy did not rise during the male migrations, and existing regional variations persisted unchanged.[34] Even in Sicily, where concerns about female purity were especially intense, peasants seemed much less concerned about lapses in female morality than middle-class observers. In Sambuca di Sicilia, several local women bore numerous illegitimate children during their husbands' absences yet they remained married, and they gave birth to legitimate children again after their husbands returned. Only one case of infanticide occurred in this town during the years of mass migration (when over half of the town's 10,000 residents applied for permission to emigrate).[35] Prostitution, a cynical farmer reported from Basilicata in 1909, was rare, "because of the shortage of men."[36] An observer of a later Italian migration reached the conclusion that "overpopulation is not conducive to trysts."[37]

While moralists worried about sexuality, the white widows — like their sons and husbands abroad — worried about heavy burdens of work. The most important economic niche for Italian women in the global economy remained that of subsistence producer of food and clothing, working without wages for family consumption. Visitors from Europe and America noted with disgust the heavy and exhausting work of peasant women in Italy. In the North, they found women plowing, haying, carrying heavy loads of hay, wood, stone, and water, grinding their own grains on hand mills, making their own bread, raising small animals, harvesting, gleaning, and processing food.[38] In the Val D'Aosta, a mountainous area of northwestern Italy where most men migrated seasonally to France, the investigator for the Jacini commission reported women were "the true beasts of burden" of the region.[39]

In the center and South, too, observers noted, "the women work like slaves," carrying as many as 70 kilograms of produce on their heads, often only eight days after giving birth.[40] Emma Ciccotosto, the daughter of a peasant family in the Abruzzi, whose father was absent for most of her childhood reported:

> Everything my mother knew she learned from her mother, and that included a lot of farm work . . . everything was done by hand. Our farm grew corn, wheat, olives, and flax, as well as vegetables and poultry. We had two sheep which my mother milked. . . . Occasionally we raised a

> pig which my mother would sell . . . all the animals were stabled at the far end of the house and their manure was thrown out of a small door into the yard where it piled up until we had time to collect it to spread on the ground for the next season's planting.[41]

"We" meant this woman and her young children, aided by groups of neighbors working communally at harvest times.

Census takers listed one in three Italian women as active workers in 1901, although women's work rates had declined, especially in the South, from over 50 percent over the previous thirty years. From Naples in 1909, Oreste Bordiga reported that female day laborers actually outnumbered male in one region of heavy male emigration.[42] Farther south, fewer families had secure tenure of land for food production. In Sicily competition for agricultural jobs had so intensified after unification that men replaced women even in their traditional jobs as seasonal harvesters of olives and nuts. The most desperate women in such regions asserted feudal rights to collect snails and greens on uncultivated lands.[43] "They graze like animals," a middle-class observer noted with horror. Still, he sympathized more with such efforts than reporters from Rome who found mendicant women, "scarcely recognizable as human," begging together with their "masses of semi-naked children."[44] The luckiest white widows in southern areas with few work options for women turned to their parents, in-laws, and female neighbors for support when necessary. ("Your true relatives are your neighbors," a Sicilian proverb reminded them.[45]) They raised small household animals to eat or sell, and they opened tiny shops, trying to earn a few extra cents as "penny capitalists." They responded not only to a local culture that increasingly condemned women who worked in the fields as immoral but also to proverbs that harshly criticized as worthless women who did not work. They did not want to appear as female "brooms" that swept away family resources through their idleness.[46]

Determined to avoid the appearance of laziness, Sicilian mothers trained daughters to sit all day at their spindles, needlework, or weaving, even as domestic production of cloth declined irregularly but precipitously between 1870 and 1900. Spinning and weaving remained a chore on Emma Ciccotosto's mother's long list of tasks on her Abruzzi farm, too. And in Lombardy and Piedmont, domestic production continued alongside both silk- or wool-spinning cottage industries and the modern spinning and weaving mills that developed there after 1880.[47] In many of Italy's other regions, women had begun to put aside their spindles and looms already in the 1870s, as cheap imported cloth rendered their labors unprofitable.[48]

Even the hard work of white widows sometimes violated bourgeois notions of appropriate female behavior, just as work in far-away fields or construction camps rendered peasant men uncivilized. Middle-class men reported that peasant women aged quickly and they were ready to criticize men who mourned their dead mules more than they did their dead wives.[49] For women and men alike, migration threatened to reinforce the already powerful stigmas of rural life and the already stark differences of Italy's two races.

Still, it would be a mistake to overstate the continuity in white widows' lives. Because they had to rely on letters and to pay cash to have them written, migration spurred illiterate women to learn to read and write. After scarcely changing for forty years, female literacy rates jumped upward in Sicily after 1900. With men's departure, women also crossed the permeable boundary separating family from public life. A few, fearing abandonment, petitioned the national government in Rome to find their husbands. Many more entered local city halls with their midwives to register officially the births of their children, whose fathers, the local clerks noted, were "in America."[50] Although entrenched in their paese, women's participation in international family economies forced them to confront Italy's national state. Their family concerns made them, too, interested in the wider world.

International family economies

International family economies were the most visible expression of transnationalism as a working-class way of life among migrants. Families accepted separations to keep their members as fully employed as possible, guaranteeing family security. But men and women also schemed how and when to reunite or (in the case of the unmarried) to marry. Whether to locate a family's "home base"[51] in Italy or abroad was a complex decision, emerging from a gradual process of learning about wages and prices in several places. While we cannot trace the learning process itself with any precision, the factors influencing families' decisions are clear enough.

Migration required families to learn what levels of security and material well-being they obtained by combining cash earnings with food-raising and cloth production at home. They learned to translate the cash in dollars, milreis, pesos, crowns, pounds, francs, or marks into lire, and to compare prices of land, housing, food, and clothing in several currencies on several continents. Families also assessed the moral worth of differing work for men, women, and children in a variety of communities at home and abroad. In locating their family home base, they considered both the financial linkage of work and consumption and the quality of work, family, and community attainable in many nodes of their village-based diasporas.

Central to their decision-making was learning the real value of cash earnings. Letters home regularly declared men's earnings, and then explained what the local currency meant if transformed into the moneta, scudi or lire, of Italy. In Europe and North America, unskilled sojourners typically earned wages a third lower than those of natives of the country where they worked. In the U.S., that meant about $1.50 daily at the turn of the century or — given seasonal unemployment in construction — a bit less than $300 yearly.[52] Sojourners cared little that natives earned more, for their own wages seemed generous once translated into lire, and compared to the wages of rural Italy. In the Abruzzi in the 1870s, for example, agricultural workers earned 1–2 lire a day (except during harvests, when wages rose to 15 lire). Peasant men averaged 25 to 50 lire in income monthly but worked only nine or ten months yearly. By contrast, in the Americas, a male sojourner working in construction earned 117 lire a month in the spring and 170 in the autumn. Even if a migrant man borrowed 100 lire to pay his passage, and paid interest of 80 to 100 additional lire to a padrone, or suffered two or three months' unemployment, foreign wages were clearly higher than in Italy.[53] A yearly income of $300 was more than five times that of a prosperous peasant family in Italy's North, and even more when compared to a struggling and landless peasant family there or in the southern regions of Italy.

Unfortunately for sojourners, living costs abroad were also much higher than in Italy. Observers who found life in the all-male camps and boarding houses appalling also always noted (but rarely praised) the men's thriftiness. Most men struggled to keep expenses below $200 yearly. When necessary, sojourners denied themselves food, sex, and pleasure to generate a surplus. Natives, in turn, fumed over migrants' "unAmerican" living standards. They recognized that Italian and Chinese workers could live on their low wages because they ate little more than rice, pasta, bread, and vegetables while native Canadians, Americans, and Argentines wanted meat, and lots of it. In fact, native workers also needed high wages in order to pay for the higher costs of meat, housing, reproduction, and family security in developing economies. While Italian men, too, had dependants, they chose to earn where wages were high in order to spend them on reproduction where prices were low — in Italy.[54]

In Italy, peasant cash incomes of 350 to 500 lire a year — combined with considerable subsistence production of food and cloth — could sustain life and ensure the reproduction of the next generation. Although Italy's peasants lived very poorly, died young, and complained of "la miseria" (misery) they rarely starved. Surplus cash from abroad thus opened possibilities for greater security, comfort, and pleasure. Already in 1880, a postal clerk in Picinisco saw almost 130,000 lire cross his counter — or about 150 lire for each of the 850 Europe-bound migrants from his small town.[55] Men from the Abruzzi later claimed that they could save 1000 to 1500 lire during a good year's work in the U.S. A year of work in most of Europe generated about 600 lire in savings, and six months' work in Germany produced a surplus of 300 to 500 lire. Even that more modest sum equaled the annual cash income of a peasant family.[56] American savings struck peasants as extraordinary windfalls.

International family economies developed out of rational, shrewd choices. They allowed men to earn where wages were relatively high and to spend where prices were relatively low. Not surprisingly, remittances to Italy soared from 13 million lire in 1861 to 127 million lire in 1880 and then to 254 million yearly after 1890 and 846 million yearly after 1906. So large was the cash inflow that it ended Italy's negative balance of foreign trade by 1912. Emigration had become one of Italy's largest industries.[57] Still, men's remittances purchased security and comfort at home only if women and children there continued to produce food and clothing. The occasional woman who strove for bourgeois standards and attempted to live in "idleness" on remittances — as some clearly did — fell victim to local gossip and charges of ozio (laziness).[58]

The decision to spend wages in Italy was rational in yet another way. Male sojourners could not easily support a family transplanted to the countries where they worked at wages so much lower than those of natives. In New York, where the typical Italian man earned $250–$350 yearly in the early twentieth century, Robert Chapin estimated that a family of four to six required about $800 a year to live adequately.[59] In Buenos Aires, the discrepancy between an unskilled laborer's earnings and estimated costs of family living was equally large.[60] Settlement abroad was possible only when a man could find year-round or better-paid work, when a family could count on additional sources of cash income, or when it could continue subsistence production of food and clothing.

Thus, "THE JOB . . . this blood-sucking thing" loomed large in decisions about the location of a family home base, and in women's and children's migrations. Only the pampas of Argentina and the plantations of Brazil and Australia offered migrants possibilities to work together as families, combining wage-earning and subsistence production. Cities offered women better wage-earning opportunities — in domestic service and industry — but only if they worked independently and away from their families. Families reached different conclusions when faced with such choices. Sicily had the highest rates of female migration, in part because women could better contribute to family economies abroad than at home. Still, Piedmont — where women worked in both subsistence production and textile factories — also had high rates of female migration, mainly to higher-wage textile towns like Lyons or Paterson, New Jersey (see Table 3.4).

Young people also viewed work options like these from their own unique perspectives. While economists noted how male departures raised wages on local labor markets, younger people also noticed their effects on local marriage markets. Although married men predominated among Italy's migrants, many young men also migrated once or more often prior to marriage.[61] Parents of sons wondered if they might fail to return or send remittances if tempted by women while abroad; parents of daughters worried about finding husbands for them. Parental fears certainly fueled the close surveillance of young men in work camps, about which the young man cited above complained. They also fueled changes in marriage customs. In places like Pisticci and Sambuca in the South, parents responded to male migration by transferring more land and houses to their daughters as dowries.[62] In Sicily, folklorist Giuseppe Pitrè reported that a propertyless young man could marry "with no more than the pants he wore" — in acknowledgment of his wage-earning potential abroad.[63] The transfer of property to women did not increase female economic power. It instead reflected women's disadvantaged position on an unbalanced marriage market. Parents replaced dowries of bedsheets and trousseaux with houses and land in order to lure back young men unsure they wanted to "far fortuna" in Italy.

For married couples, by contrast, maintaining a home base in the village often required repeated migrations and repeated separations.

In the North, contract laborers might go every year for six months to other European countries. In the South, too, Emma Ciccotosto's father left her Abruzzi home twice for short sojourns in the U.S. before emigrating again to Australia in the 1920s. Only after fifteen years of circulatory migrations to two continents did he call for his wife and children to establish a new home base in Australia. In the lives of the Ciccotosto family one senses uncertainty, false starts, and mixed reactions, not only to the hardships of separation, but to prospects for achieving family security in Italy, America, and Australia. A description of consumption and community ties in the paese and its diaspora makes the complex decision-making in international family economies particularly poignant. Whether families maintained a home base in the paese, or transferred it to the diaspora, they had to balance rewards with sacrifices.

"Americans" in paese: villages of return

The fact that 50 percent of Italy's migrants returned to their country around the turn of the century suggests intense commitment to old and familiar ways of life. Indeed, scholars have usually emphasized that male migrants were at most "conservative adventurers."[64] Temporary migration certainly propped up subsistence production and traditional gender relations at home. It limited the inroads of individual wage-earning on family work groups and it prevented the total proletarianization of the workers of the world. Young men gained some autonomy during migration but expectations of loyalty and mutual aid within family and kin groups persisted with the migrants' return. At the same time, however, returners rapidly demonstrated that they intended to live in new ways in their hometowns. They became eager consumers of the relatively narrow range of modern consumer goods available in rural Italy, and they abandoned many ancient symbolic expressions of deference to local elites. The paese had created its diaspora, but the diaspora in turn transformed the paese. By the end of the mass migrations, Italy's two "races" neither looked nor lived as differently as they had in the past.

Many men migrated from Italy with hopes of establishing their families on a more secure footing within existing paese class hierarchies — as landed peasants, rentiers, or small artisans and shopkeepers. These represented the occupations most respected in much of rural Italy in the nineteenth century, and they defined a more comfortable and civilized way of life within reach of Italy's peasants.[65]

Few scholars have systematically measured upward mobility among returners.[66] But existing evidence suggests that occupational mobility was limited, occurring largely among agriculturalists, as wage-earning braccianti and peasants' sons acquired leases or small plots of land. Few returners became wealthy landowners. Their remittances had too quickly inflated the value of local lands in those regions where any land at all was available for sale (as it was not everywhere). Dino Cinel has argued that remittances and return were highest where landholding was common but fragmented, so that land remained widely available in relatively small and inexpensive parcels.[67] Government investigators found migration had almost no impact on landholding patterns in southern areas of highly concentrated ownership and large estates. At most, a few owners of latifundia (large estates) leased land parcels under more favorable conditions, or substituted leasing for gang labor, creating "mini-fundia." Observers agreed that migration introduced few agricultural innovations, at least in the South.[68] Migration did not substitute for land reform; land hunger among Italy's peasants remained intense and produced a new wave of land occupations and social conflict after World War I.[69]

By contrast, many returners purchased new housing or improved the houses they owned, guaranteeing a healthier, more comfortable, and more secure if modest life for many families. By expanding their houses, migrants visibly affirmed their commitment to their home communities and to their family's continued presence there.[70] They signaled their improved social status, too, since the very poorest had no home, and sometimes still lived as mendicants or in caves. Investment in housing was especially common where sojourners returned from the Americas. In the area around Salerno in the first decade of the century, peasants paid 1000 lire (one to three years' savings) for houses government investigators found to be "substantial." These were free-standing, two-story dwellings, quite unlike the older one- and two-story attached houses of most villages. With plastered walls and floors of brick, they had proper windows, a separate kitchen, an internal staircase, and two upstairs rooms. Some had balconies and shutters, sturdy wooden doors, and a tiled roof.[71] In some villages, returners introduced acetylene lighting or made inquiries about obtaining electrical service in towns that still lacked piped water or sewage. The contrast of the homes of migrants to the "old black homes, piled on top of each other and separated only by torturous, dark little streets" impressed even moralizing middle-class observers.[72]

Surprisingly few migrants invested in small businesses, and when they did, the results were not always positive. A Sicilian who worked in Cairo for twenty years, first as a street vendor, and then in a small shop, succeeded after twenty years in returning home with sufficient capital to open a shop in his hometown of Termini Immerese. He also built a small house, and purchased a tiny parcel of land, which he leased to a peasant, hoping to live like a proper rentier. He complained to a government investigator that "with the high price of [the peasant's] labor I failed to make both ends meet, so I shall have to emigrate again."[73] Remittances and savings invested in housing did raise local demand for artisans in the building trades but migration's overall impact on trade and commerce in Italy seems limited. In Sambuca di Sicilia, where about 40 percent of peasants returned, more arranged for children to marry artisans and small businessmen than to open new businesses themselves. Scholars of later return migrations to the South agree that temporary migrations financed purchase of modern consumer durables, not commercial or industrial development.[74]

Cash earned abroad transformed consumption habits among rural dwellers, allowing many to enjoy a considerable sense of material improvement. Already in the 1870s and early 1880s, the investigators for the Jacini commission had noticed with displeasure peasants' new-found fascination with purchasing from the local pizzicagnolo, or small shopkeeper who sold cheap trinkets, cloth, household equipment, and pipes. They castigated peasants' desires for simple pleasures as threatening decadence and they urged on peasants the proper resignation to poverty that Christian teaching demanded.[75] In many regions of the North and a few of the larger cities of the South, poor men and women had already abandoned traditional costumes before migration. Bourgeois observers complained that "homewoven cloth does not sufficiently satisfy [women's] vanity," and that women's "mania for dressing up" increased daily.[76] Young men returning from military service had even learned to perfume themselves, they claimed. Mass migration accelerated a transformation in consumption already underway. Poor peasants abandoned sandals and wooden clogs for shoes; men purchased suits of wool and velvet, hats, and watches. By 1910, traditional women's costumes had disappeared in the coastal areas of Campania, and even in isolated mountain villages women donned only distinctive scarves, shawls, or headwear. Peasants enjoyed purchasing urban clothing that gave them the civilized look of city dwellers.

Cash also allowed poorer migrants to eat better, opening the possibility that their children could eventually grow as tall and robust as urbanites. Investigators for the Jacini commission had established that Italian peasants consumed sufficient food to survive while working hard but that many failed to meet draft requirements for stature. They also concluded that peasant diets — regardless of region — were monotonous, largely vegetarian, and probably nutritionally insufficient. (Debates about the cause of pellagra in the North especially pointed to dietary problems.)[77] Northern peasants ate as much as a kilo of corn meal polenta daily; their evening meal was a thick soup (minestra) of grains, legumes, and vegetables, along with a little wine and occasionally cheese. Everywhere in Italy, breads of mixed grains were more common among the poor than wheat loaves. Farther south, peasants ate more bread than corn meal, and more wheat, both as bread and (around Naples) as pasta — a holiday treat. Southerners ate more greens, fruits, and vegetables, but their only seasoning was oil. Southern peasants rarely ate cheese or meat, and even on holidays they drank wine mixed with water (vinello).[78]

Twenty-five years later government investigators found sweeping changes in the diet of southern peasants. Wheat bread increasingly replaced other grains. Peasants more often kept a pig, and they used its meat to season pasta dishes weekly rather than twice a year. Local elites complained even about these changes. From the Molise, one wrote of inflation resulting when "the wives of the Americans arrive at the marketplace and buy up all the fresh fish newly arrived from Temoli, regardless of price."[79] Migration and spreading commercial agriculture (which introduced new crops, and raised output of other food products) allowed peasants to create what we now recognize as the popular regional cuisines of Italy.[80] Peasants around Vasto developed a taste for hot peppers, while enthusiasm for maccheroni spread throughout the South. In the North, peasants substituted more rice, and wheat bread, for the ubiquitous polenta, and they began eating more meat (pork, rabbit, chicken) and cheese.[81]

Returners' purchases suggest that migration had been a strategy for improving the security and material well-being of individuals and families; it was not a strategy for transforming the paese itself. Only occasionally did migrants invest their cash savings in communal improvements. In southern Italy, an occasional returner founded a hospital or built a seaside walkway; visitors more frequently reported contributions to popular religious festivals with foreign remittances.[82] Nevertheless, migration did have an influence on the paese by altering the tenor of local social relations.

Change was most noticeable among the young and male, and had begun years earlier, among draftees. Investigators for the Jacini commission reported young men rebelling against their fathers by choosing their own brides. Contrasting the restrained severity of older courting customs, one investigator in the area around Lanciano reported with distress that young men wanted to "be in command of their own hearts." Serenading in the evening had become a common element in courtship throughout central and some parts of southern Italy by the 1890s. It signaled the spread of romantic love's challenge to parental control of arranged marriages. Where parents attempted to maintain control, their obstreperous sons eloped with their lovers, who (in a rare display of female rebelliousness) pleaded abduction and rape, forcing the parents to accept a marriage.[83] Twenty-five years later, young men even insisted on marrying before migrating; they wanted to earn wages for their wives, they insisted, not for their parents.[84]

Men claimed increased autonomy beyond the family, too. Bourgeois observers in Italy noted that returners, like the army veterans, resented the deference expected by local landowners. The Jacini commission investigators faithfully recorded local complaints about disrespectful young soldiers unable to find happiness working under traditional contracts, with few rewards or pleasures. Later, from the Abruzzi, a government investigator noted that migrated men, too, acquired "a consciousness of their dignity as men, and a sense of independence and liberty when face to face with the class of gentlemen." Those same gentlemen, of course, experienced returners as arrogant, and mocked their "exalted dreams."[85] From Puglia, another investigator agreed that the migrant man returned "radically transformed," "energetic, conscious of his dignity and more confident in his strengths, no longer convinced he was born only to serve."[86]

This same investigator, Errico Presutti, reported that returners had formed a consumer cooperative. Agricultural strikes increased in southern and northern Italy during the 1890s and again during the first decade of the twentieth century, as both migration and return peaked. Rates of labor militancy in Sicily were actually highest in towns with high rates of migration.[87] If migration had ever functioned as a safety valve alleviating tensions between urban rich and rural poor, it was not a particularly effective one. Such expressions of discontent were exceptional, but the more common and symbolic declarations of independence were just as threatening to local landowners. A popular ditty among peasants mocked young gentlemen, telling them to remove their gloves, put their umbrellas aside, and prepare to work in the fields — because the peasants were off to America.[88] Returners refused to doff hats, kiss hands, or otherwise offer traditional symbols of servility.

Even their peasant neighbors recognized the returners as "new men."[89] In Basilicata and Calabria, Ernesto Marenghi reported that the fellow villagers called the new, modern houses of the returners the "case americane" (American houses). The same was true in the Abruzzi and Molise where returned migrants built American-style houses on the outskirts of their home villages, in new quarters inhabited almost exclusively by other former migrants. Fellow villagers called the returners, too, "Americans." In some towns the "Americans" not only lived in their own new quarter, but they intermarried among themselves. Even returners from Europe sometimes found themselves dubbed "Germans," "Swiss," or "French." In Friuli, where many returned from Germany, the "germanesi" brought with them new beer- and schnapps-drinking habits that marked their leisure hours as distinctive.[90] Much more than new consumer desires marked the returned migrants as "americani" or "germanesi." Their experience in the wider world, and the autonomy and confidence it produced, made them seem like Americans or Germans at home — still comfortable in their paese but no longer entirely of it.

Life in the "colonies"

By moving their home bases abroad, migrant families could accomplish many of the same ends as the returners. The colonies formed abroad by Italy's many villages were strikingly diverse.[91] In New York, migrants from Italy crowded into two- and three-room apartments in six-story tenements on the Lower East Side and in Harlem, where most of their neighbors were migrants from other parts of Italy. In San Francisco's North Beach, in Buenos Aires, and in Marseilles, by contrast, they rented space in smaller multi-family dwellings and counted natives and immigrants of other backgrounds among their neighbors. In mining and textile communities in France, and in Tampa's cigar-making district of Ybor City, migrants lived in ramshackle and small but frequently private dwellings. In Pittsburgh, the soot of the steel mills fell on their heads; in Montreal, migrants complained of devastating winter cold and in Queensland of oppressive winter heat. In Lugano and Trieste, migrants' native neighbors themselves spoke dialects of Italian. In New York and Buenos Aires, furthermore, the accents of Palermo, Naples, Bari, and Turin intermingled. In Stuttgart migrants heard Schwäbisch (the native Swabian dialect), and in the Ruhr the Polish accents of transplants from Germany's eastern empire. In Paris, on Chicago's North Side, or New York's Elizabeth Street, migrants' push-carts and shops offered familiar goods. In southern Brazil or northern Colorado, by contrast, a company or plantation store offered the only wares, often in an unknown tongue.[92]

Despite this diversity, the relocated family bases of migrants shared some common features around the world. Life abroad freed migrants from the arrogance of the Italian elite but also required fundamental changes to family relations, especially between the sexes and the generations. Like returners, unskilled migrants rarely experienced rapid upward mobility by going abroad. Seasonal unemployment among men remained too common. Furthermore, most found that all members of the family, including women and children, had to work for cash in order to pay the higher costs of living. With families enmeshed in a capitalist consumer economy, migrants in cities like Buenos Aires, New York, or Marseilles seemed to enjoy more, and more exciting, choices as consumers, especially if their households contained skilled or white-collar workers or petty merchants. But dedication to family security abroad prevented the majority of unskilled migrants from indulging too many consumer desires — at least if we assume that evidence from the U.S. is typical for other places.[93] In the U.S., the search for family security was arguably more daunting than in a village of return. The choice to settle abroad was not clearly the most financially rational or emotionally satisfying one for the least skilled, low-income migrants in the years before World War I. That so many returned is scarcely surprising.

Escape from the humiliations of poverty and from local elites figured prominently in migrants' explanations for relocating abroad. "Hey! Mr. Sir," one peasant called tauntingly to a local landowner in the early twentieth century: "better dead in America than alive in Italy!" Another explained more simply that "in Italy we lived like beasts," whereas in the diaspora, a man could more easily feel like a civilized human being.[94] Few of Italy's landowning rural elite chose a life abroad during the years of the mass migrations. And when they did, there was no hand-kissing or hat-doffing in the diaspora.

Escape from local elites did not mean immediate economic prosperity. Male wages in the diaspora were simply too low and male unemployment too common to support a secure or comfortable family life. In families of unskilled men, women and children had to become wage earners if a family was to survive. According to Samuel Baily, 38 percent of married Italian women in selected districts in Buenos Aires in 1895 worked for wages outside the home, while 60 to 80 percent of unmarried women did so. In New York, ten years later, about 7 percent of married women worked outside the home, and almost half of unmarried women did so. Women in Buenos Aires may have been more willing to leave home to work because they found employment in small-scale shops owned by fellow migrants, or because more were from Italy's North. Southern women in New York, by contrast, had to move well beyond family circles to work for wages in garment manufacturing owned by Jewish, German, or American employers.[95] Still, even they found ways to contribute cash to family coffers.

In New York, married women often found industrial work to do at home. As homeworkers, they sewed pants, cracked nuts, and made artificial flowers, effectively combining wage earning with domestic chores.[96] Only a few New York census takers counted homeworkers as wage earners, but those who did found rates of wage-earning among married and unmarried Italian women much like those of Buenos Aires.[97] Although less well-studied, Italian families in France also seemed to expect some form of wage-earning from married women and children alike.[98]

In Buenos Aires and in New York, immigrant women transformed domestic work into cash by taking into their homes sojourners separated from their families but no longer dependent on padroni. Family-based boarding was a new social relationship, and a source of income without precedent in Italy. About a third of New York's Italian wives kept boarders at any one time; in Buenos Aires, the proportion was over 40 percent. Boarders included relatives as well as friends and neighbors who paid small fees in exchange for meals, a bed, and clean laundry. Boarders' payments made only small contributions to family incomes — scarcely a third of what women could earn by working outside the home — but they allowed many married women to continue working within the family circle.[99]

Even with women and children working at home production or "keeping" boarders, families of unskilled workers in New York and Buenos Aires struggled to pay high housing, food, and clothing costs. Robert Chapin estimated that in New York in 1909 three-quarters of Italian immigrant families earned less than the $800 needed for a family of four to six to live securely. He reported migrants from Italy spending a fifth of their income on rent, almost half on food, and 12 percent on clothing. Yet, surprisingly, 58 percent of the Italian families he surveyed also reported a budgetary surplus and some savings.[100] In Buenos Aires, where information about family budgets is scantier, Italian incomes were also well below those reformers found necessary to maintain minimum standards, yet over half of all working-class families there too reported some savings.[101] Many immigrant families obviously began saving at relatively low incomes, forgoing the pleasures of consumption.

How did they do it? In neither New York nor Buenos Aires did they stint on food. Immigrant families in New York ate more eggs, meat, cheese, and milk than returners did. Many reported the satisfaction they felt in eating great quantities. "Don't you remember how our paesani here in America ate to their hearts' delight till they were belching like pigs and how they dumped mountains of uneaten food out of the window?" one woman in New York reminisced with pleasurable exaggeration.[102] In Buenos Aires, too, where beef raised on the nearby pampas was plentiful, immigrants remembered eating well.[103] More than any other consumer choice, plentiful food and drink symbolized well-being for transplanted migrants in both North and South America.

Evidence from New York suggests that families more often pinched pennies on clothing, housing, recreation, and entertainment. Married women (and their youngest children) sometimes still went barefoot, made their own clothes, and remodeled cast-offs. Men claimed only very small pleasures for themselves — notably tobacco, a coffee, or a beer at a nearby café. Husbands and sons turned over their paychecks to their wives and mothers, who were the budget managers for most families.[104] Adolescent children proved the least willing to forgo the pleasures of urban consumerism. Some openly resented having to turn over "every cent"; others quietly "borrowed from mother." One girl in Rhode Island reported "I'll never forget the time I got my first pay . . . I went downtown, first, and I spent a lot, more than half of my money . . . I just went hog wild." More complied and "handed our pays in." Girls desired modern clothing, urban shoes, and hats. Boys wanted to enjoy sports, shows, dancehalls, or other commercial entertainments.[105] Like their age-mates in Italy, too, young men and women abroad wanted to choose their own spouses based on mutual attraction and romantic love. Clothing and commercial entertainment were part of new rituals of American courtship.[106] If their parents enjoyed a sense of escape from local elites, children found freedom in consumption, romance, and marriage.

Women continued to manage family efforts at saving, even as their working lives changed significantly. The women's power over the purse in Sicily had justified their control over consumption and savings and their choice of the marriage partners of their children.[107] Somewhat reluctantly, mothers in New York relinquished control over their sons' wages, and many boys essentially became boarders in their own families. Having already accepted that their daughters left the family circle to work for wages, however, southern Italian mothers seemed less willing to allow adolescent girls control over their own money, leisure time, or romantic life. As one girl concluded wearily, her recreational life was limited to "going up and coming down" between the apartment and her tenement building's "stoop" (front steps).[108] Reported an immigrant mother in Pittsburgh, "Josephine was not allowed to go out with a boy . . . but with the second girl, no. She would go out with the boyfriend. Times change."[109]

In both densely settled New York and lower-density Buenos Aires, immigrant mothers ruthlessly limited the amount they spent on rent in order to facilitate saving. In New York, families moved frequently in search of the cheapest quarters. Many families also willingly lived in very tight accommodations, with families of five or more in two rooms. In New York's downtown settlements, the youngest families with one or two children doubled up. They squeezed three families into a four-room apartment with a shared kitchen where boarders also slept. Census takers called these "partner households." Because so many Italian immigrants crowded into small, old tenement dwellings, social reformers like Robert Chapin despaired of their ever learning or accepting American housing standards.[110] Their desire to save — an otherwise highly valued "Yankee" characteristic — kept them from becoming American consumers.

Mothers' parsimony — like that of the men without women — originated in part in their desire to assist family members still in the paese. It also reflected persisting feelings of economic insecurity. While family solidarity facilitated migration, transplantation also demanded changes at the center of family life. Families that relocated abroad were peasants and artisans who had only recently severed their ties to subsistence production. They no longer produced for their own consumption and needed cash for all their needs. They had relinquished their ideal of family work groups in order to benefit from women's and children's cash contributions to a consumer family economy. And they faced the growing independence of their own children as wage earners, consumers, and eager seekers of marital romance. In the colonies of Italy's many diasporas, the workers of the world more nearly resembled the proletarians Marx had imagined them to be. It heightened, rather than diminished, their desire for family security.

Efforts to find security took predictable forms — subsistence production, friendship, and mutual assistance among paesani. Wherever they could, Italian immigrants made heroic efforts to grow some of their own food; some even raised rabbits on fire escapes and goats in tenement basements.[111] Home ownership was even more important, having been relatively common in the homeland the migrants had left, even among quite poor peasants. Returners could purchase houses with savings from a year or two of work by one successful male migrant. In the Americas, by contrast, home ownership required a long-term struggle, one in which the wages of growing children were crucial — but disputed — elements. Few migrant families in high-density New York became home owners before the 1920s; a government investigation in 1909 found home ownership rates of only 1 percent. By contrast, in Buenos Aires, with its lower-scale housing, and its higher proportions of skilled and white-collar workers, 16 percent of Italian families in 1904 already lived in houses they owned. Compared to returners' successes, however, even these seemed shabby results.[112]

Understandably, in both Buenos Aires and New York, migrants settled near relatives, and near fellow villagers. Because they formed such a large part of the Buenos Aires population, migrants of Italy could scarcely segregate themselves from those of different backgrounds in this burgeoning city. In New York, the self-segregation of migrants — a relatively much smaller group in this multi-ethnic city — was far more noticeable. In Lower Manhattan, some blocks housed Sicilians and others Neapolitans. The Sicilian village of Cinisi had its own colony farther up the East Side. In 1911 migrants from Sicily's Termini Immerese sent two photographers home to record the procession of the local patron saint (Saint Agostino Novello) and to bring the results back for display in New York.[113] Kinship, friendship, and neighborliness rooted in the village remained important sources of social and financial security for migrants living on insecure incomes far from home.

In both the villages of return and the satellites of village-based diasporas families sought security. But while returners seemed determined to cast aside old mores and customs, families abroad more often appeared as social, moral, and fiscal conservatives. They struggled to reproduce at least some of the customs of the villages they had left behind. The search for security in Italy's "colonies" reinforced migrants' ties to family and paesani. But it also inevitably pushed workers of the world beyond the social relationships that had served them well before and during migration. Some turned to familiar institutions like the Catholic church to provide a sense of security and belonging. Others saw security in diaspora nationalism, and in new forms of cooperation with migrants from their home villages or from elsewhere in Italy. Others — as Marx had predicted — sought security through class-based activism. The proletarian cosmopolitanism of Italy's many diasporas did not prevent — and in some ways even encouraged — migrant workers to become particularly creative practitioners of working-class internationalism.

5

NATIONALISM AND INTERNATIONALISM IN ITALY'S PROLETARIAN DIASPORAS, 1870–1914

"For us there are no frontiers."[1]

The epigraph above provides yet another apt summary of life in Italy's proletarian diasporas. Yet "for us there are no frontiers" was not a comment on the transnational lives of men without women or white widows. These were the words of a nameless anarchist from Italy, on trial in London in 1894. Like this anarchist, almost all socialists in the nineteenth century and communists in the twentieth century proclaimed themselves internationalists.[2] In a sentiment best expressed by Marx's 1848 call in the *Communist Manifesto* — "workers of the world, unite" — radicals of the era of the mass migration assumed workers would ignore national differences as they sought justice and economic equality. Few of the labor internationalists of the era of mass migrations had attempted to theorize about migration, but had they done so they might have logically expected all migrants to support labor internationalism, and make it victorious.[3]

They were wrong. Labor internationalism disappeared rather quickly as a major force in world history. The most influential labor movements of the nineteenth and twentieth centuries were national ones.[4] In addition, for many "workers of the world," ties to kin (familism), fellow villagers (localism), and religious faith proved as salient as class-based experiences of capitalist exploitation in the work place. And, although few of Italy's migrants had strong national identities when they left home, many more developed them as they lived and worked abroad.

Still, in the years between 1870 and 1914, labor internationalism was a major force in world history, and it posed a considerable challenge to nationalism. It was just a radical fantasy without relevance for ordinary men in Italy's diasporas. In fact, labor migrants — usually led by exiles among them — explored more varieties of internationalism than Marx himself had imagined. On several continents, migrants from Italy pioneered in the development of multiethnic, multinational, and transnational organizing strategies.[5] Some labor activist exiles (like those of the Risorgimento) also formed their own distinctive diasporas, allowing us — for example — to speak of Italian anarchism as a transnational ideology unbound by migration and spreading wherever Italy's anarchists went.

Labor internationalism was an important strategy explored by migrants in search of economic and social security but only as one of many alternatives. Migrants turned also to national labor movements, to a Catholic church with its own universalism and history of international organization, and to mutual aid among those who shared a common birthplace in the home village. The relative importance of class, nationalism, internationalism, region, and religion varied enormously in the "other Italies" that gradually coalesced from the satellites of Italy's many village-based diasporas.[6] The influence of labor internationalism, regionalism, and Catholicism became particularly obvious in fierce competitions among the minority of literate migrants to define these other Italies as unified communities of Italians and to lead them in a unified direction.[7] In France, labor activists often became leaders, while in Argentina, cross-class ties allowed both labor internationalism and diaspora nationalism with close ties to Italy's consuls to flourish in an Italian ethnic enclave. In New York, the missionaries of a Catholic church still estranged from the Italian state more successfully unified migrants than did the middle-class prominenti who glorified civiltà italiana. Everywhere, competition among varieties of diaspora nationalism, paese- and kin-centered regionalism, and exiles' internationalist ideologies was quite pronounced until World War I.

Proletarian internationalism; proletarian nationalism

The internationalism of earlier patriots like Giuseppe Garibaldi and Giuseppe Mazzini had been the midwife for Italy's labor movement, as Chapter 2 showed. In 1870, Garibaldi openly proclaimed himself a supporter of both the Paris Commune and the International Workingmen's Association (the First International), founded in 1864. As the Hero of Two Worlds withdrew from military action, his garibaldini carried on his military and internationalist initiatives.

Men like Francesco Nullo organized soldiers from Italy to fight for the national liberation of Poland in 1863, while Amilcare Cipriani and others went with Garibaldi's son to fight against Turkey in Serbia and Greece in 1897. Much later, men like Lorenzo Vanelli, Cesare Ravera, and Ilio Barontini fought in an Italian Garibaldi brigade in the Spanish Civil War (see p. 150 below).[8]

The aging Giuseppe Mazzini, by contrast, was reluctant to embrace class struggle as a principle, although he (more than Garibaldi) otherwise refused to compromise with the new Italian monarchy. In Italy, Mazzini's followers organized workers' societies focused on education, cooperation, and mutual aid in Italy's northern cities and towns. Although the Mazzinians gradually lost leadership in Italy's emerging labor movement, mutual aid societies spread throughout Italy's North and center in the 1860s and 1870s and in the South in the 1880s.[9] After 1900, Catholic priests equally hostile to class conflict determined to bridge the gap between Italy's two races by encouraging mutual aid societies that united small landowners and small businessmen with poorer workers.[10]

As the split between Garibaldi and Mazzini revealed, the meaning of internationalism also changed as it became associated with anarchism, communism, socialism, and dreams of proletarian revolution. Along with the better-known leaders of the Russian, French, and German anarchist and socialist movements, men like Saverio Friscia in South Italy and Andrea Costa in the North joined the First International in the aftermath of Italian unification.[11] By the 1880s, anarchists and socialists had created their own separate revolutionary movements, and Socialists in Europe formed their own (Second) International in Paris in 1889. Tensions between differing theories of working-class activism plagued the development of labor movements around the world for the next eighty years.

As the ideology of internationalism changed, so did its proponents. Only a handful of returned Risorgimento exiles became internationalists; most of those were garibaldini of humble origin. Typical of Italy's newer labor internationalists were men too young to have participated in the Risorgimento, like Enrico Bignami who was born in 1844 to poor parents. Having gained a rudimentary education in his hometown of Lodi, Bignami fought briefly with Garibaldi against Austria at age twenty-two. A Mazzinian republican, he then founded a newspaper for workers, *Il proletario*. In 1870, he became an internationalist. He later became attracted to evolutionary socialism and the ideas of Benoît Malon. A founder of the Partito Operaio Italiano (P.O.I., Italy's first labor party, organized by former anarchist Andrea Costa), Bignami never rejected free thought, and he remained sympathetic to anarchism even after becoming a socialist. After fighting on the barricades in the May, 1898 "fatti di maggio" revolts in Milan, Bignami fled into permanent exile in Switzerland.[12]

He was not alone. Between 1870 and 1900, the Italian government, still smarting from its battles with southern brigands, turned its new security forces to the threat of anarchist and socialist internationalists. It jailed left-wing activists like Bignami and banned or shut down their newspapers and workers' societies. Each wave of police repression pushed out a new generation of exiles. The biographies collected in Franco Andreucci and Tommaso Detti's *Il movimento operaio* ("The Labor Movement") reveal that over a third of Italy's most prominent prewar labor activists fled into exile one or more times. Except for a handful, all were men. Anarchist exiles were 57 percent in the 1870s, 63 percent in the 1880s, and 21 percent in the 1890s. During the 1890s and early 1900s, socialist exiles increased rapidly to 74 percent.[13] Like the exiles of the Risorgimento, almost 90 percent of Italy's exiles eventually returned home.

As the lives of Bignami and other activists reveal, Italy's labor movement began as a coalition of socially mobile middle- and lower-middle-class men (many of them intellectuals) and urban artisans. Viewing peasant wage earners and industrial workers as an exploited majority, they tried to build organizations that would draw Italy's disparaged rural workers into the nation and into national politics. An overwhelmingly rural country, where peasants still vastly outnumbered industrial workers, Italy's labor movement acquired a distinctive national character.[14]

Its first defining characteristic was its preference for anarchism and insurrection rooted in community solidarity. The second was its hostility to the state and to electoral politics. One labor historian has even claimed that Italy's labor movement was "born anarchist" in the 1860s, when Bakunin lived in Sicily and Naples for a time.[15] In the 1860s and 1870s anarchists from Italy hoped to build an insurrectionary movement to destroy state, church, and elite hegemony.[16] During the years of brigandage and peasant revolt following unification, early anarchists like Florido Matteucci (who helped organize an abortive insurrection in the southern province of Benevento in 1877) saw peasants as potential revolutionaries.[17] The anti-draft and food riots, tax revolts, and land occupations of peasants were often violent — and women participated in them along with men.[18] This restiveness of Italy's rural majority attracted Italy's anarchists, even as they diverged into contending groups that favored propaganda of the deed (terrorism), Luigi Galleani's anti-organizational anarchism, or Errico Malatesta's anarcho-syndicalism.[19] Their hopes for revolution seem less absurd when we remember, as Louise Tilly argued, that "the serious possibility of revolution opened up at least once in every decade from 1860 to the Fascist seizure of power in 1922."[20]

Hostility to the state shaped the development of Italy's labor unions and its Socialist Party. The first unions of industrial workers formed in the 1880s, mainly in the North. They, along with the P.O.I., stressed workers' autonomy (operaismo) and the creation of local chambers of labor, not electoral action.[21] These were male initiatives; women of the popular classes later joined unions in small numbers but fraternalism excluded them from mutual aid societies as the lack of franchise excluded them from electoral politics.[22]

Neither Marxist socialists nor the later theorists of German Social Democracy provided obvious blueprints for Italy's socialists.[23] The first group emphasized too exclusively the industrial proletariat as the engine of revolutionary change; Italy scarcely had modern factories before 1900. The second group instead pinned their hopes on enfranchised workers, educated by labor activism, capturing power over the state. The strategy was scarcely imaginable in Italy, where no workers at all voted before 1882, and where artisans and petty shopkeepers formed a minuscule portion of the electorate until 1912.

Italy's Socialist Party (P.S.I.), modeled on Germany's Social Democrats, formed in 1893 as a new round of peasant revolts brought down the Italian government. When peasant leagues joined the new Socialist Party en masse, Sicilian insurrectionists became the largest group of party members.[24] None was a voter. A few years later, members of the peasant leagues that struck repeatedly in the Po Valley were the largest group of P.S.I. members.[25] The party also attracted members in early metal and textile unions in the North, along with a considerable number of middle-class voters.[26] Even as the Socialist Party grew, however, insurrection continued in the fields and in the piazzas of growing cities. Massive revolts North and South in 1898, including the fatti di maggio, remind us that Italy's citizens sought power at work and in the piazza, not within the state.[27] Within the P.S.I., revolutionaries often outnumbered reformers.[28]

The onset of mass migration did not end workers' protests or organizing. On the contrary, the number of strikes and strikers increased during the peak years of emigration 1900–14.[29] Unwilling to use repression like his predecessor Francesco Crispi in the 1890s, the new liberal Prime Minister Giovanni Giolitti sought new solutions, and nudged his government to action. With the cooperation of small numbers of socialist legislators, Giolitti's supporters passed laws to regulate the working conditions of women and children. The government also now recognized workers' rights to organize and to strike. Syndicalists, revolutionary socialists, and anarcho-syndicalists continued to reject collaboration with his liberal government, but Italy's Socialist Party argued for extending the franchise and their own legislative power. In 1913, the first year when a sizeable number of ordinary men could vote, the Socialist Party had more members than any other Italian party, and one-quarter of the deputies elected were socialists.[30]

The influence of the Italian left thus expanded modestly during the peak years of emigration from Italy. By 1912, the country had three competing national federations of unionized workers (socialists, syndicalists, and Catholic activists). The largest was the union of agricultural workers. Only 12 percent of Italy's industrial workers and 5 percent of agriculturalists were union members in 1914, and only a few hundred thousand could vote before 1913. Membership in mutual aid societies was higher but not by much. Nevertheless, a movement born anarchist and internationalist in the 1860s now claimed influence, both directly and indirectly, at the national level.

Most migrants who left Italy to become workers of the world had never heard the word internationalism. They had never voted; most had never belonged to a labor union or workers' party. Some — although certainly not the majority — had joined a workers' friendly society (società di mutuo soccorso). Many more migrants had direct or indirect personal experience with the rebellions and strikes that had racked Italy's countryside since the 1860s. Migration was not a strategy of the passive or an alternative to worker protest. Sicilians from towns with peasant protests in the 1860s and 1890s, for example, migrated at higher rates than other Sicilians and continued to do so into the twentieth century when a new wave of agricultural strikes swept their home villages.[31] Migrants from northern towns like Biella were no strangers to workplace conflicts or strikes, nor were the seasonal migrants to Europe from the day laborer families of the Po Valley or the America-bound migrants among Sicily's wheat-raising braccianti.[32] Although it is impossible to describe such migrants ideologically, some very preliminary evidence on rank-and-file activists, collected by the Italian police, suggests that anarchists predominated.[33] Overall, the history of Italy's labor movement shows common people nurturing a continued antagonism toward their national state.

With their limited experience of labor militancy, soon humble migrants again faced the complex interaction of labor nationalism and internationalism abroad. It was a violent encounter. Stigmatized as rebellious criminals at home and as padrone slaves abroad, migrants rarely found much welcome. Native workers accused them of stealing their jobs, breaking their strikes, depressing their salaries and leaving them unemployed. In Europe, North America, and Australia, Italy's laborers encountered frequent protests, violence, and calls for their exclusion by native-born workers.[34] Periodically, more bloody confrontations occurred. Mobs composed mostly of workers went on anti-Italian rampages in Lyons, Marseilles, Grenoble, Berne, Zurich, and through several mining towns in Australia and the United States.[35] The best-known of these anti-Italian riots was the Aigues-Mortes massacre of 1893. Singing the "Marseillaise" and marching behind red flags, French workers hunted down Italians, leaving an undetermined number dead and scores injured.[36] Like the Chinese, Italy's migrants first exposed, and then challenged, the prevalence of what Robert Paris has called "proletarian nationalism" in labor movements around the world.[37]

National labor movements initially sought to protect native workers by excluding migrants. Australia's movement provides the clearest case of working-class nativism. Anglo-Celts feared migrant workers as economic competitors. The Australian labor movement maintained that the immigration of Italians (referred to as the "olive peril" or as "swarthy") threatened to "erupt and spread like lava," undermining the wages of superior white workers.[38] During the 1880s, French socialists also advocated immigration restrictions and limits on foreigners' access to jobs.[39] Similarly, Swiss socialists consented to state repression of Italian strikers, radicals, and exiled activists.[40] In the U.S., the American Federation of Labor demanded first the exclusion of contract labor and then (after achieving that in 1885) reductions in overall immigration.[41]

Some native workers feared migrants' activism more than their passivity, however. Even sojourners displayed the restiveness and willingness to revolt characteristic of seasonal unskilled workers and peasants in Italy. Both in the U.S. and in Canada, unskilled sojourners struck spontaneously without eliciting a positive response from native unions.[42] In the U.S., skilled Italians excluded from American trade unions founded by earlier German and Irish immigrants formed Italian unions instead.[43] In Australia, migrant miners were a militant presence already in the 1892 mining strike at Broken Hill; in 1905 migrant lumberjacks supported striking Australian workers and endorsed Labor Party candidates.[44] Still, migrant militancy alone provided insufficient grounds for establishing good working relations between migrants and national labor movements. Those working relationships developed as a new generation of Italy's exiled internationalists first challenged proletarian nationalism and then forged connections between the migrants and national labor movements around the world.

The meanings of labor internationalism

The organizing strategies and successes of exiled internationalists varied considerably in the other Italies. In a small way, the variation reflected anarchists' and socialists' preferred destinations. In the 1870s, prominent anarchist exiles like Carlo Cafiero, Tito Zanardelli, and even Errico Malatesta traveled almost exclusively within Europe. After 1880 (as police restrictions against them tightened) almost half — the prominent activists Errico Malatesta and Francesco Saverio Merlino among them — instead headed for North and South America.[45] By contrast, 80 percent of the socialists fleeing Italy after the repression of the fasci siciliani and the fatti di maggio in the 1890s went to Europe. Differences in labor initiatives also reflected the nature of the national labor movements Italy's exiles encountered abroad.[46] In Europe, socialist exiles found labor movements of socialist (Germany, Austria, Switzerland) or syndicalist (France) leanings. In Latin America, anarchist exiles found no labor movements at all. In the English-speaking world, anarchist exiles encountered labor reform movements dominated by reformers hostile to radicalism and revolution, and comfortable working within capitalism and within political systems where workers already voted. Anarchist exiles hostile to state and unionization — Luigi Galleani is a good example — could scarcely expect to find a comfortable place in the American Federation of Labor (A.F.L.) or the German Socialist Party.[47]

Italy's socialist exiles in Europe — men like Luigi Campolonghi, Angiolo Cabrini, and Giacinto Menotti Serrati — were among the earliest critics of proletarian nationalism. Already in 1893 in Zurich, Antonio Labriola had exhorted his comrades of the Second

International to welcome migrant Italians into their unions.[48] As exiles eager to end Italian scabbing and to "win" the migrants for socialism, exiles asked native workers, as did one migrant in France,

> Who are we? Men like you, like you oppressed, exploited, unhappy. Like you we have capitalists that starve us and governments that bayonet us. . . . Brothers of France, if there is war, it should not be between us, unhappy slaves that we are.[49]

Socialist labor movements in Europe quickly responded. Joint effort between exiled socialists and their native comrades led to the formation of an Italian Workers' Club in Vienna in 1894.[50] The following year in Switzerland, exile and local labor organizations joined in establishing the Unione Socialista Italiana.[51] In 1898, in an effort to encourage migrant membership, the German construction workers' union began publishing its own newspaper for migrants, *L'operaio italiano*.[52] Internationalists all, the organizers never questioned the fact that the migrants were in fact "Italians."

Support for foreign initiatives came also from Italy's Società Umanitaria (Humanitarian Society). The socialist reformers of Milan had founded the society in 1882 to promote workers' education. In the 1890s it began to collect statistics about unemployment and foreign demand for labor. Unhappy with the P.S.I.'s decision to make local labor chambers and trade federations responsible for organizing migrants, Umanitaria began its own campaign to "conquer migration" for socialism.[53] It provided travelers' aid at railroad stations and border crossings, and opened labor offices in regions of intensive emigration and in European cities with many migrant laborers.[54] German, Austrian, and Swiss unions asked Umanitaria to provide Italian-speaking organizers during strikes. During the winter months, Umanitaria representatives toured regions of mass emigration with foreign labor organizers, extolling unionization and international labor solidarity.[55]

Umanitaria's work with foreign labor movements highlighted the problems of migrants in a world of national labor movements. In Italy, the P.S.I. approved a resolution to expel members failing to join unions while abroad.[56] The P.S.I. and Umanitaria urged foreign unions to accept migrants inscribed in unions in Italy without additional fees; a number of Swiss and German construction unions effectively became international organizations by doing so.[57] Unresolved was the problem of the peasant worker, who might belong to the Italian agricultural workers' union (Federterra) but who worked abroad in the construction industry, facing double dues. Eventually, an International Secretariat of (national) labor centers attempted to grapple with organizational problems like these.[58] By 1910, Italian membership in European unions had risen substantially.

For the socialists in the Second International, internationalism thus meant multinational cooperation in the work place. The Second International supported free migration across borders and the unionization of workers without regard to citizenship. Socialists did not attack national notions of citizenship, however, or demand voting rights for migrant workers; they expected migrant workers to return home rather than acquire citizenship or political rights in the countries where they worked. Thus, the socialist parties of Germany and Switzerland showed little interest in the "foreign workers" (as they called them) from Italy. They were not potential citizens or voters but rather potential voters for Italy's P.S.I.[59]

In sharp contrast to Germany and Switzerland, Latin America and France, like the U.S., viewed migrants as prospective citizens. In these countries, internationalism more often meant cooperation within a national labor movement formed by peoples of differing backgrounds and cultures. Today — at least in English — we would call such organizations multi-ethnic; labor militants at the time, however (especially if they spoke French, Portuguese, Spanish, or Italian), called them international.[60] Multi-ethnic labor organizations assumed two quite different forms. Some were unitary (or creole) groups that mixed together workers of many cultural backgrounds. Others were culturally plural with each group organized separately within a multi-ethnic federation.

Where Italy's anarchists (and, later, anarcho-syndicalists) settled among French, Portuguese, and Spanish workers, labor movements most often took the form of multi-ethnic creoles. Beginning at the turn of the century, exiles like Luigi Campolonghi in France independently led scores of Italian workers to join with their French co-workers in local unions. Jointly, they participated in strikes of construction and dock workers in Marseilles and of steelworkers in Meurthe-et-Moselle.[61] With some notable exceptions, France's Confédération Général du Travail (C.G.T.) insisted that "for us, union workers, there exist no nations . . . there exists only one nation — that of the exploited class."[62] Italian anarchists and anarcho-syndicalists also gained prominence in creole labor movements in Marseilles, Tunis,[63] Argentina, São Paulo (Brazil), Tampa, and in the North American Industrial Workers of the World (I.W.W.). Their organizations explicitly referred to their multi-ethnic membership in labeling themselves cosmopolitan or international, as did the Argentine bakers' union, the Societad Cosmopolita de Resistencia (Cosmopolitan Resistance Society). In Europe, socialists recognized multi-ethnic labor experiments as a form of internationalism. They especially praised the I.W.W. and Argentina's labor federations for uniting workers of diverse backgrounds.[64]

Unlike France, Latin America and North Africa had no significant working classes or labor movements until the migrants created them, usually along syndicalist or anarcho-syndicalist lines. Crucial to anarchists' successes in Argentina was the influence of Errico Malatesta and Pietro Gori, who lived in the country from 1885 to 1889 and from 1898 to 1902 respectively. Malatesta spread his new ideas about unionization through multi-lingual publications and extensive lecture tours around the world. By the turn of the century, under Gori's influence, and with the assistance of lesser-known exiles like Napoleon Papini and Eugenio Pellaco, Argentine anarchism had extended its activities beyond artisan circles. Exiles were instrumental in organizing many of the first railroad, textile, and meat-packing unions among Argentina's emerging urban and migrant proletariat. In 1901, unions united in Argentina's first labor federation, the Federación Obrera Regional Argentina (F.O.R.A.). Over half the delegates to the convention had Italian surnames and the activists met in La Boca. Within a year, F.O.R.A. became an anarchist organization and syndicalists established a rival federation, the Union General del Trabajo (U.G.T.). Overall, about a quarter of Italian workers in Argentina's small-scale industries were members of one group or the other, and another quarter supported their strikes.[65]

Italian immigrants also played a central role in building the smaller labor movement of Brazil. In Brazil in the 1870s and 1880s, Italian radicals Eugenio Sartori, Giovanni Rossi, and others had formed short-lived rural cooperatives and agricultural communes, including the much-studied anarchist Cecilia Colony in the state of Parana.[66] Beginning in the mid-1890s radicals like Gigi Damiani, Antonio Picarolo, Vincenzo Vacirca, and Edmondo Rossoni instead worked in São Paulo in an emerging working class dominated by Italian migrants. They published scores of papers and established small but militant labor unions. Italian radical influence was particularly visible in the Brazilian Workers' Federation, which maintained its anarcho-syndicalist character from its founding in 1906 until 1920. Syndicalist hostility to political action thus characterized the national labor movements Italy's migrants created (Italy, Brazil, Argentina) and the French movement in which they so actively participated.[67]

Where exiles from Italy held ideological views sharply different from the existing national labor movements, internationalism more often involved segmentation along cultural lines. Ethnic segmentation was especially pervasive in the U.S. The American Socialist Party offered membership in segregated foreign-language sections.[68] The mainstream, labor-reform unionists of Samuel Gomper's A.F.L. only reluctantly opened its doors to unskilled immigrant workers in a few industrial unions that were themselves exceptions within the trade-union-dominated A.F.L. Garment workers' unions — the International Ladies' Garment Workers' Union (I.L.G.W.U., a member of the A.F.L.), and later the Amalgamated Clothing Workers of America (A.C.W.A.) — both welcomed Italian migrants and some — the Sicilian Bellanca brothers, Vincenzo Vacirca, Luigi Antonini — became prominent leaders. The I.L.G.W.U. and A.C.W.A. accepted ethnic segmentation when migrants demanded such autonomy (in Italian locals 89 and 48). Italians' requests for ethnic autonomy reflected their early discomfort in unions initially dominated by Yiddish speakers.[69]

Because garment work was an important niche for women workers from Italy, female labor activism first emerged in this industry. In Buenos Aires, Italian women workers began striking during the first decade of the century. In New York, by comparison, organizers of the huge strike, the "uprising of the 20,000," in 1909 complained about the passivity and fearfulness of Italian women workers, especially when compared to their Jewish counterparts. In later strikes (1913, 1919), however, Italian women both joined the I.L.G.W.U. and became picket-line militants. They even had their own Italian-language newspaper, *L'operaia* ("The Woman Worker"). The first female labor leaders in the diaspora — women like Angela Bambace — emerged from the I.L.G.W.U. and A.C.W.A. By the 1930s, Italian women outnumbered men in locals 48 and 89. They were also more assertive in claiming leadership and representation in the larger union. While women were otherwise invisible among internationalist exiles in the other Italies, women's work as wage earners quickly transformed a few female immigrants into activists.[70]

Ethnic segmentation seemed in keeping with the A.F.L.'s opposition to socialist internationalism. But even the more radical French garment makers' unions also tolerated ethnic locals for a time in the early twentieth century.[71] And even the revolutionary and syndicalist I.W.W. — the A.F.L.'s main competitor — with its leadership of prominent Italian leaders, like Carlo Tresca, Joseph Ettor, and Arturo Giovanitti, accepted some segmentation. The I.W.W. strike committee in Lawrence, Mass., in 1912, for example, included representatives of fourteen immigrant communities of strikers even though the I.W.W. followed French syndicalists in emphasizing that all workers, regardless of background, belonged in "one big industrial union."[72] The first person killed in Lawrence had been a young Italian worker, Anna LoPizzo.[73] Women (many from Biella) were also prominent in a later strike of silk workers in Paterson, New Jersey.

In Europe, the multinational empire of Austria and the multilingual nation of Switzerland also experimented with ethnic segmentation in their labor movements, encouraging the spread of national identities. Before World War I, Austria had three Italian-speaking provinces — Trento, parts of Vorarlberg, and Süd Tirol (Alto Adige) — and the port city of Trieste. All three areas attracted labor migrants from Italy — called regnicoli ("subjects of the king"). In 1898, when Austrian socialists chose to segment along national lines into Italian-, German-, and Slovenian-speaking branches, Italian syndicalists in multi-ethnic Trieste objected. Proclaiming themselves internationalist opponents of the socialists' national chauvinism, they insisted that native-born Slovenians and Italians could unite with the regnicoli in creole unions.[74]

In Switzerland's Italian-speaking Ticino province, similar conflicts occurred between native-born Italian socialist reformers and revolutionary migrants. Regnicoli migrants were a quarter of Ticino's workers in the late nineteenth century. During Milan's fatti di maggio revolts in 1898, Italian masons and laborers thronged Swiss roads, heading for home "to fight the war against the monarchy." Initially joined in the Italian Socialist Union of Switzerland (U.S.I.S.), native-born and regnicoli split after only three years. As social democrats, the native Italian Swiss formed their own political party (the U.S.L.I. or Italian Socialist Workers' Party) but encouraged migrants to join the German-dominated and socialist Gewerkschaftsbund (union federation) with which they cooperated. Italian exile regnicoli, however, preferred a revolutionary socialist party more like the P.S.I., and in fact called themselves the Italian

Socialist Party in Switzerland (P.S.I.S.).[75] In Austria's Trentino, newspaper editor, regnicolo and labor migrant Benito Mussolini made much the same argument.[76] Revolutionary socialist exiles wanted to organize Italians abroad to support revolutionary change at home. Theirs was a transnational rather than a multi-ethnic internationalism.

Among Italy's anarchists, too, internationalism frequently meant transnational solidarity, communication, and support among Italy's migrants around the world.[77] No other group of exiles was more in need of such networks of support. Propaganda of the deed and the migrant anarchists (Sante Caserio, Michele Angiolillo, and Gaetano Bresci) who assassinated European political leaders in 1890–1900 heightened fears of anarchism originating in Italy.[78] International police efforts to identify anarchists, to prevent them from crossing borders, and to keep them under surveillance forced anarchist exiles onto particularly long and complex paths of exile.[79] Errico Malatesta lived and worked in Austria, Switzerland, Argentina, France, and the U.S. before he settled into fifteen years of stable exile in England. As an organizer of multi-ethnic unions, Malatesta enjoyed contacts with activists of many backgrounds. But the more typical, modest, and impoverished Italian anarchist — along with migrants wedded to propaganda of the deed or individualism — worked more often within an Italian-speaking world.[80]

Global networks of Italian anarchists in the 1870s and 1880s reveal themselves in the spread of Italian-language anarchist publications to Egypt, Australia, Switzerland, France, Argentina, and the United States. Wherever two or three exiles gathered, they published a paper and began propaganda meetings. In this self-contained world, exiles nevertheless pursued international themes with considerable persistence. Their newspapers and clubs carried news of workers around the world, of clerical atrocities in Italy and abroad, and of the repeated failure of Italian consuls to protect ordinary laborers. Articles that first appeared in France, Italy, or Buenos Aires reappeared in Paterson, London, and Lucerne.[81] Anarchists in Brooklyn celebrated the Mexican revolution, while deploring Italy's war with Turkey, and mourning the death of Francisco Ferrer in Spain.[82]

Transnationalism was not internationalism as the socialists of the Second International envisioned or practiced it. Like ethnic segmentation in multi-ethnic but national labor movements, transnationalism carried within it the makings of diaspora nationalism. It was probably no accident that in both Trieste and Trento,

Italian-speaking native workers, through their interaction with syndicalist regnicoli, eventually became strong supporters of incorporation into Italy. Failure to possess multi-ethnic Trieste, in particular, would become a symbol of honor betrayed as Italian nationalists saw it in the years after World War I. Still, transnational organizing did not always generate diaspora nationalism. No nationalist movement comparable to Trento's or Trieste's developed in Switzerland. And Italy's anarchists remained far too hostile to state power for their transnational support networks ever to serve nationalist ends. In the era of the mass migration, as in the early nineteenth century, many forms of labor internationalism remained locked in a curious embrace — half symbiotic, half hostile — with national labor movements. The same could be said of migrants in the other Italies as internationalists battled nationalists wanting to "make" Italians, Frenchmen, Americans, Argentines, and good Catholics out of the workers of the world.

Altre Italie: creating other Italies

Over time, the satellites of Italy's many village-based diasporas merged almost imperceptibly to become "other Italies." None yet had particularly close ties to Italy's state (a topic addressed in greater detail in Chapter 6). In differing cities, Catholic and labor activists, representatives of regional associations, and policies of the receiving countries encouraged this transformation, creating "other Italies" characterized by quite diverse forms of diaspora nationalism. Almost everywhere, Catholic, regional, and labor leaders struggled fiercely with each other for dominance of "little Italies" forming on foreign soil. Complicating their work was the fact that the migrants who had relocated their home bases to the other Italies remained skeptical of organized movements of almost all types; they remained focused on family and neighborhood as the best guarantors of economic security. Like peasants in Italy, the workers of the world vacillated between periods of intense and rebellious communal solidarity — "almost religious in quality," observers noted — and the pursuit of family security through constant, individual, migration in search of work.[83]

Three examples must suffice to illustrate the diversity of the "other Italies." In Tampa, Florida — as in many French cities in the 1920s (see Chapter 6) — labor activists successfully defined the Italian community as a bastion of labor internationalism and anti-clericalism. In Argentina, liberal, republican, and middle-class migrants dominated as leaders, producing a pluralist diaspora nationalism that embraced, rather than foundered on, class differences among migrants. In New York, battles between internationalist exiles and nationalist prominenti enamored of civiltà italiana continued until World War I. Paese-based diasporas and the continuation of regionalism prevented even the illusion of a unified community there. At most a shared (if not always practiced) Catholic religious faith united migrants from many regional backgrounds in New York. But in the years before World War I — when the Catholic church had not yet dropped its opposition to Italy's "godless state" — few priests claimed to represent Italians as a national group, even when they worked in "national" parishes among other Italians.

The multi-ethnic cigar-making community of Ybor City, just outside Tampa, Florida, in the U.S., was in many ways part of Latin America. Its multi-ethnic population and its militant anarcho-syndicalist labor movement resembled that of Buenos Aires or São Paulo more than New York. Ybor City's workers had three large mutual aid societies — one for Italians (mainly Sicilians), one for Cubans, and one for Asturians (from Spain). Each national society had a magnificent building, and offered a wide range of services to its members — libraries, dining and meeting rooms, ballrooms, and the services of a doctor paid by insurance taken out by members. Ybor City's labor movement, by contrast, was a multi-ethnic creole based on a shared commitment to syndicalism. Sicilians, Asturians, and Cubans organized, bargained, and struck (repeatedly) as members of a single multi-ethnic industrial union they described as "cosmopolitan." Their "lector" who read to them as they worked in local cigar factories chose from the literatures and radical publications of Spain, France, Italy, England, and the U.S. Sicilians in Ybor City often learned Spanish before English. During strikes, workers congregated at any of the mutual aid societies. While men led and spoke for the strikers, women struck along with them. During more peaceful times, Sicilians, Spaniards, and Cubans also mixed for recreational activities in their fraternal societies. Ybor City's small businessmen also enjoyed close familial and commercial ties to the migrant militants. But it was the radical and anti-clerical activists — like the Sicilians Alfonso Coniglio and Giovanni Vaccaro — who spoke for Ybor City Italians — not businessmen or priests.[84]

Farther south, Argentina's Buenos Aires had a much larger migrant middle class, one that had played an important role in the development of diaspora nationalism since the days of Garibaldi. In Buenos Aires, artisans and businessmen of the Risorgimento had founded the oldest mutual aid societies and Italian-language newspapers already in the 1850s. After successfully contributing to the overthrow of the authoritarian dictator Rosas (see Chapter 2), many of the founders of Italian community organizations became supporters of Bartólome Mitre. (Mitre was first governor of Buenos Aires as an independent province, and then President of a reunited Argentina in the 1860s.) Although monarchists sometimes still battled with republicans for leadership of Buenos Aires' community institutions after unification, both groups emphasized national, Italian, unity among all migrants and worked quite closely with Italy's consuls. The mutual aid societies founded by the Risorgimento exiles were old, large, and prosperous by the time the workers of the world arrived.

These newer mass migrations threatened the hegemony of older institutions. Between 1880 and 1900, dozens of new, small, and paese-based mutual societies — local and regional in orientation — challenged the older republican and monarchist Italian national societies. Over time, the national societies recovered and grew because they offered so many services the new migrants desired. They ran Italian-language schools (which may have educated 40 percent of Italian students in a city where public education scarcely existed before 1905) and a hospital open to all Italians.[85] Ultimately about 30 percent of the male, and 10 percent of the female, population of Italian Buenos Aires belonged to one of the city's mutual aid societies. Somewhat more (25 to 50 percent) also moved within circles influenced by the labor internationalists, as we saw above.

The Catholic church, by contrast, seemed of limited influence in Italian Buenos Aires. Argentina was a Catholic country, and the Vatican saw few special needs for Italian migrants where they were in no danger of converting to Protestantism. A shared Catholicism linked immigrants with the natives, rather than becoming an element of their sense of cultural distinctiveness. In addition, the Catholic orders (mainly Salesians) that ministered to migrants in Argentina faced a community where the anti-clerical and Masonic commitments of the republican Risorgimento, and the new Italian state, continued to influence the mutual aid societies and newspapers well into the twentieth century.[86]

Despite the fact that Italian workers sometimes worked for Italian employers, class conflicts were not particularly intense in Italian Buenos Aires. Certainly, anarchists and syndicalists railed against capitalists, and they struck against the largest Italian manufacturers of the city. But most Italian employers were small-scale producers or artisans themselves, and not a few were sympathetic to the new ideologies of socialism, anarchism, and syndicalism; many of the rest were republicans. Thus, employment in small, Italian-owned shops did not reproduce the hostile relations of Italy's landowning elites and peasants.[87] On the contrary, the older mutual aid societies and the newspapers noted with concern the special problems of workers.[88] The newspaper *La patria* grew rapidly during the mass migrations and became the third or fourth largest newspaper in Argentina. It reported on older community institutions, newer mutual societies, and labor initiatives alike, and it railed at the Argentine government for its failure to address and to solve the excesses of class conflict.[89] According to Samuel Baily, the Italian-origin middle-class leaders of Buenos Aires were more public-spirited and more concerned with the well-being of the entire national group than were Italy's elites in the same years.[90] They were not political brokers, however, since electoral politics had not yet become a mass phenomenon in Argentina. "Italian" Buenos Aires was secular, well-organized, and united around a common sense of Italian national identity crossing class lines. Buenos Aires' "other Italy" claimed to have built the country of Argentina through its labors but it had no possibility of ruling it politically.

In New York, by contrast, no institutions of the (very small) exile migrations of the Risorgimento survived to greet the mass migrants. With few opportunities in New York for middle-class or white-collar workers from Italy, it was also a settlement sharply divided by class difference. New York's prominenti were a tiny and isolated group of recently arrived professionals (doctors, lawyers, music teachers) and a somewhat larger (but also small) group of upwardly mobile padroni, labor agents, successful businessmen, and bankers. Few (besides the padroni and labor agents) employed significant numbers of Italian workers, and even the padroni — like the professionals — competed with each other to build the networks of clients they needed to survive economically.[91] All aspired to leadership as ethnic brokers who could connect Italian migrants to middle-class America, and especially — as did banker and labor contractor Louis V. Fugazy — to the political machines of American cities. In the same way, New York's labor internationalists aspired to be "radical ethnic brokers," linking workers to national and international labor movements in the U.S. and elsewhere.[92]

Neither group enjoyed much success at first. Seeking economic security, New York's new migrants instead formed hundreds of new mutual aid societies. Most were extremely small, with limited financial resources. Almost all were fraternal societies that limited membership to men from a particular town or region. (Women found a place in immigrant fraternalism and mutualism, often in auxiliary societies, only as these organizations began to lose membership after 1910.) Padroni transformed into businessmen founded not a few of these village-based societies; they often sold the insurance such groups provided for members. While mutual aid societies provided crucial forms of financial assistance to poor immigrants, the small clubs of New York began confederating only after 1910.[93] However, U.S. policy toward Italy's migrants inadvertently reinforced regional differences among them. U.S. policy was influenced by Italian intellectuals' thinking on the racial difference between northern and southern Italians (and especially on the criminal propensity of the latter). Beginning in 1899 and continuing until 1921, the U.S. recorded and analyzed northern and southern Italians separately in its immigration statistics.

Just as New York's mutual aid societies reflected the hundreds of village-based diasporas in the city, its Italian-language newspapers reflected competition among those who desired to lead a united Italian community. Neither *L'eco*, *Il progresso*, or *Corriere della Sera* became as influential, or as widely read, as *La patria* in Argentina. All defended the good name of Italy from an onslaught of American criticism of migrants viewed as padrone slaves, instruments for depressing wages, and criminals. But editors like Felice Tocci and Carlo Barsotti battled with each other mainly over issues of personal honor and prestige, not over the political past or future of the homeland.[94] Neither was a particularly persuasive proponent of a nationalism tied to the modern state of Italy; nor did either enjoy particularly close ties to Italy's consuls. The one successful long-term campaign of New York's prominenti erected a statue to Christopher Columbus — a symbol of the early influence of civiltà italiana, not the glories of modern Italy. As nationalists still concerned with spreading civiltà italiana, New York's newspaper editors had little tolerance for labor activists or internationalists. Only the *Corriere della Sera* sometimes reported on strikes or labor meetings.[95]

Prominenti's battles with labor activists nevertheless repeatedly disrupted everyday business within New York's many mutual aid societies. These groups sponsored regular banquets where men ate together and toasted each other and their social solidarity. Radical members might suddenly unfurl a red flag in the middle of such a banquet or loudly burst into revolutionary song. The evolution of the Order of the Sons of Italy (a first national federation of mutual aid societies founded in Philadelphia in 1911 which grew rapidly after World War I) reflected the competition of middle- and working-class men for leadership of the regional and paese-based societies. The initial by-laws of the Sons of Italy promised to censure any member who scabbed during a strike; by the 1920s, however, many of the federation's members were middle-class enthusiasts of fascism.[96]

New York's Italian labor movement remained divided among anarchists, syndicalists, and socialists and between supporters of the I.L.G.W.U. or A.C.W.A. and the I.W.W. The history of the Italian radicals of New York has not yet been told; it remains buried in dozens of ephemeral radical newspapers. Small groups of radicals active within one or more mutual aid societies from their home region did not quickly coalesce into a city-wide Italian-speaking radical movement. Thus in Brooklyn, a small group of Sicilian syndicalists had their own newspaper where they read about international events and about strikes of Italian shoemakers and barbers (mainly organized by the I.W.W.). The group generated its own small radical counter-culture with May Day spring festivals and "anarchist baptisms" (where they named their children Alba (Dawn) or, later, Lenin). Only in the garment makers' unions, and in the years of the I.W.W.'s Lawrence and Paterson strikes, did radical Italian labor activists in New York seem to speak for a sizeable number of migrants. And, again in later years, New York's Italians would repeatedly elect the radical (yet Republican) Vito Marcantonio to represent the large Italian-origin settlement in East Harlem.[97]

In New York, neither labor internationalists nor prominent diaspora nationalists could claim to speak for the city's 350,000 Italian residents with much authority. New York remained a collection of family- and paese-based social networks. Yet these alone, apparently, could not create a sense of security and belonging in a fragmented and chaotic city. In a theme familiar to the historians of business and everyday life in southern Italy, migrants in New York sought guarantees of protection from the dangers of the world beyond their immediate social circles. In small ways, they recreated the thuggery, extortionism, and criminality their critics claimed to find in the Sicilian mafia and the Calabrian camorra. Already in 1890, New York newspapers highlighted horrific stories of bombings by an Italian Black Hand society. From the migrants' perspective, there was no criminal conspiracy or Black Hand behind protection rackets — just a desire for order and security and a handful of isolated toughs willing to sell their services promising order in a multi-ethnic city. For most young men on New York's streets, friends were more important forms of protection than professional toughs. Recalled one New York migrant of the Lower East Side, "Look, if you didn't organize your friends for your own protection you were dead . . . sure we had gangs. There was the Elizabeth Street gang, the Mulberry Street gang. That's all there was, gangs."[98]

In the kaleidoscopic, and apparently somewhat threatening, world of New York, the Catholic church provided an element of familiarity and security. Catholic priests played a much more important role in New York than in Buenos Aires for only they could claim to represent an institution that united all migrants from Italy. In Rome, the Vatican was aware of the special problems and dangers facing migrants, especially in Protestant countries like the U.S., and it established both a Prelate of Italian Emigration and a College of Italian Emigration that prepared priests for missionary work among Italy's migrants. Catholic priests were important forgers of Italian solidarity in the U.S. but until the twentieth century, most shared the church's hostility to Italy's state and to any form of nationalism associated with it. They, like the socialists and anarchists, were as much internationalists as nationalists in their efforts to mediate between Italy's migrants and the wider worlds around them.[99]

In addition, Catholic priests faced a migrant population that was more wedded to their popular religious practices than to financing and supporting a parish priest. Contemporaries claimed there were only 50,000 practicing Catholic Italians in New York in the 1890s and the Vatican was dismayed at the corruption and behavior it found among priests who had fled Italy in search of employment when their orders lost their means of support after unification. Even decades later, Italian migrants contributed less to their parishes than other immigrant groups; they sent their children to parochial schools in smaller numbers; they also became Protestants in larger numbers than other Catholic immigrants did.[100]

In an overwhelmingly Protestant land, Italians (like other immigrant Catholics) nevertheless quickly organized into "national," linguistically segregated, "Italian" parishes. Italian-speaking priests — whatever their opposition to the "godless state" of Italy — worked with them in large numbers. Most priests were from Italy's North; most migrants from Italy's South — a source of considerable misunderstanding and tension, especially over the observance of popular festivals. Still, the northern priests shared with their southern parishioners the disdain of an American church dominated by German and Irish bishops, and this seemed to unite them as members of one Italian nation. By 1918, there were seventy Italian parishes in New York, and most had been organized and ministered by northern Italians of the missionary orders, rather than by priests appointed by America's Catholic bishops. It was an order of immigrant nuns, headed by Mother Frances Cabrini, that in 1892 founded the Columbus Hospital that served migrant Italians in New York.[101] The contrast to Tampa — where mutual aid societies led by radicals hired a "social doctor" for their members — and to Buenos Aires — where the middle-class mutual aid societies built an Italian hospital — could not have been clearer.

A sense of Italian national identity consolidating around very different organizations and leaders thus defined the other Italies as places of quite distinctive character in the years before World War I. Diaspora nationalism in these years was not a product of the initiatives of Italy's state but of activists influenced by the ideals of civiltà italiana, the Risorgimento, the labor movement, and the church. Italian exiles and immigrant workers had successfully challenged proletarian nationalism. They had created a variety of labor initiatives that they, and their radical colleagues, understood as international. Mobile Italian exiles had become some of the most effective proselytizers of anarchist and syndicalist ideals around the world.[102] During the same years, however, new forms of nationalism had emerged in Italy's diasporas, as they did also at home in Italy. Many internationalist activists had helped to build national labor movements; some of those national labor movements tolerated segmentation along national and ethnic lines. At the same time, cross-class collaborations between middle-class and working-class Italians and between northern and southern Italians were not at all unusual or unsuccessful, especially in mutual aid societies and within the Catholic church.

Despite the efforts of church and fraternalism to unite Italians across class lines, tensions generated by class, and by the conflicting ideals of nationalism and internationalism, were visible throughout the diaspora. In the other Italies, nationalists and internationalists alike faced the challenge of earning the support of migrant workers, and of connecting them to the political parties, labor movements, and churches of either the homeland or the new countries in which they lived. By 1914, conflicts between nationalists and internationalists peaked throughout the world, with special consequences for Italy's many diasporas. With World War I, and the collapse of international labor solidarity, warring states and nationalist movements at last made firm claims on workers' loyalties. In the thirty years between the onset of World War I and the end of World War II the tension between differing definitions of Italian nationalism assumed the form of a brutal conflict. Migrants and their families found they had finally to choose national loyalties — something many had successfully avoided for several generations.

6

NATION, EMPIRE, AND DIASPORA: FASCISM AND ITS OPPONENTS

"We can say loudly and with pride that we are traitors of the Fascist patria because we are loyal to another Italy . . ."
(Carlo Rosselli)[1]

The Second International collapsed in 1914 when its constituent national labor movements chose to defend their homelands rather than labor solidarity. By 1929, Pope Pius IX had signed the Lateran Treaties, and Roman Catholicism had become the state church of an Italy it had once opposed. With the demise of labor internationalism and Catholic opposition, nationalism had triumphed in Italy and in Europe more generally. The two most important global events of the twentieth century were world wars in which workers of many countries died in large numbers to defend national interests.[2] In both wars, countries like Germany, Italy, and the United States depended on citizen soldiers to fight for the nations that few had acknowledged as their homelands 100 years earlier. In the aftermath of each war, old empires collapsed and the number of nation states in the world increased. Despite intermittent efforts to form new multinational organizations — first the League of Nations, later the United Nations — nation states had become the single most important determinants of human experience worldwide — shaping how humans lived, worked, and died.

Mass migrations, labor internationalism, and competing forms of nationalism had little place in this world of assertive, and often hostile, nation states. Indeed, wartime tensions and economic depression raised states' demand for national loyalties and conformity and heightened distrust of migratory workers. In the U.S., reformers demanded that migrants "Americanize," while in Italy, supporters of Mussolini explained "why Italians shouldn't migrate."[3] Residents of Italy, along with their families and friends abroad, faced harsh choices in the years between the two world wars. Almost thirteen million had applied to leave home in the thirty years before 1914; only four million more would do so during the next thirty years. The modest choices and comforts generated by international family economies disappeared. The interwar years brought diminished economic security for Italians — and most other peoples — wherever they lived. Ironically, the diminished security that family- and village-based diasporas could produce under these new circumstances also pushed migrants to embrace nationalism.

But in what form? Like other countries worldwide, the once new and weak nation state of Italy was more focused than in the past on creating an Italian nation willing to support it. Certainly, that is what the world's first self-proclaimed fascist insisted when he came to power in 1922, boasting that "Italy today is a world power," which expected to be treated "like a sister rather than a waitress."[4] A returned migrant and revolutionary syndicalist who had worked in Switzerland and Austria, Benito Mussolini rose to power by crushing the internationalist left to which he had once belonged; he consolidated his power by ending the long-standing separation of state from church. His elderly anarchist arch-nemesis Errico Malatesta observed rather wearily of him, "It's the same old story of the brigand who becomes the policeman."[5] As Italy's new policeman, Mussolini sought to redirect migration, sending it to the country's rural regions, to its African colonies, and to nations offering Italy advantages in exchange for its migrants' labor power.

More than any previous leader of Italy, Mussolini also sought the respect and support of Italians living abroad. He wanted them to feel part of the Italian "stirpe" (tribe or race), as he called it. Mussolini sought to transform the other Italies into demographic colonies within his fascist empire. Appealing to migrants already under considerable pressure to prove their loyalties and to conform to the nationalist expectations of the countries where they now lived, his efforts provoked considerable controversy. Among his opponents were the political dissidents (as many as 10,000) he himself chased into exile. Anti-fascists like the Rosselli brothers, Carlo, quoted above, and Nello, offered Italians at home and abroad a vision of an Italian nation unconnected to fascist empire or racism. Unlike Mussolini's, the nationalism of the anti-fascist exiles — whatever their ideological preferences — emphasized love of the patria and of liberty. In the past, the Catholic church had nurtured a worldwide religion that contrasted the godless anti-clerical government of Italy to the "real Italy" that had given birth to Catholic civilization. In the interwar years, anti-fascists created their own international movement as they urged the "true" Italians living abroad to recreate the "real," democratic Italy that Mussolini's "legal Italy" had usurped. Not since the Risorgimento had debates about nationalism, its meaning, and its practice generated so much passion in Italy's diasporas.

Hostile states and the waning of the mass migrations

Nationalism asserted itself in many forms around the world in the years prior to and just after World War I, but nowhere more obviously than in new restrictions on migration. In the nineteenth century, neither sending nor receiving countries had seriously attempted to regulate global labor markets or to control capital flows. In the twentieth century, both came to question the liberal premise that laborers, like capital, had a right to move about as they pleased. Albeit in varied ways, nations around the world saw increased surveillance of migrants and foreign workers as essential to national security. Still, the incentive to migrate in the face of climbing obstacles remained strong among Italy's citizens, and the mass migrations did not suddenly end because of restriction.

Although prospective migrants would also eventually face opposition from Italy's rulers, rising barriers in countries that had once provided work for millions were of the greatest importance. Motivations for restricting migration were complex, but often stemmed from national efforts to incorporate labor movements and workers into the national polity. In the U.S., labor had already encouraged the passage of laws excluding Chinese workers and contract workers in 1882 and 1885. In the wake of World War I, the Russian Revolution, Italy's "red biennium" of strikes and factory occupations, a wave of strikes in American steel and textile industries and during Argentina's "semana tragica" ("tragic week") all raised fears of migrant radicals. Anarchists had found themselves excluded from the U.S. (in 1902) as they were also from Argentina and Brazil (1902, 1911). All three countries deported foreign radicals at war's end and focused attention on the dangers posed by work forces of foreigners.[6]

Critics of migration in the U.S. and Canada also sought to diminish immigration totals because they feared the cultural disunity that migration produced. In the U.S., some Americanizers claimed they wanted only to give an overheated melting pot time to work its nationalizing alchemy on new arrivals. In 1917 Congress excluded illiterates as the least assimilable of immigrants; Canada followed suit in 1919. Other nativists were openly racist in opposing continued immigration. Social Darwinists like Madison Grant, and public figures like Theodore Roosevelt, claimed that native-born Anglo-Saxons or Anglo-Celts were committing race suicide by welcoming migratory hordes of inferior, mongrel peoples from Asia, and from southern, central, and eastern Europe. After excluding Chinese laborers on racial grounds in 1882, the U.S. began listing northern and southern Italians as different races (in 1899), and in 1907 it also excluded Japanese laborers. After World War I, the U.S. excluded Asians and half-closed its door to nations not already well represented in its population of settlers in 1890. After 1924, a discriminatory quota for Italians allowed only 3600 new entries from Italy each year.[7]

In Switzerland, too, fears of Überfremdung ("over-foreignization") and of political radicalism mounted after 1910 when a national census shocked Swiss citizens into recognizing that 15 percent of their resident population was foreign-born. Regulation of migration into Switzerland did not focus on reducing the numbers of border crossers, as it did in the U.S. and Canada, but rather on restricting foreigners' opportunities to work or to settle permanently. New permits for work and settlement put every foreigner under the direct surveillance of a new police agency, created specifically to oversee them. Either the restrictions or a failing economy in the 1930s worked the desired effect. By 1941, foreigners made up only 5 percent of the Swiss population.[8]

In other European countries, fears of migratory workers originated instead in the costs of constructing welfare states. Beginning with the social security provisions pioneered by Germany's anti-socialist Chancellor Bismarck in the 1870s, welfare programs had become a state mechanism for incorporating, but also taming, radical workers within the nation and its national state. Between 1880 and 1940, Germany, Austria, and France all created programs that guaranteed minimum economic and social well-being for their citizens, including workers. As unemployment benefits, old age pensions, health care, and schooling became sizeable expenses, states logically questioned why non-citizen, migrant workers should have access to them when they faced none of the obligations of citizenship, notably military service. The rise of welfare states helped to transform migrant workers into something other than neutral, cost-less, factors of production in a national economy. For the first time, they became potentially costly drains on state budgets.[9]

An Italian governmental guide to the rights of Italians living abroad in 1934 specified in considerable detail the many restrictions migrants faced when working abroad.[10] Austria and Germany granted the same unemployment, disability, and retirement rights to foreign workers as to natives, but only if the foreigners continued to live in those countries (an option precluded by high barriers against permanent residence and naturalization). Austria excluded foreigners from many professions, and from operating many businesses; it prohibited them from joining political parties or participating in meetings about public affairs; it required them to pay higher school fees for their children. Switzerland and France offered some welfare services to Italian workers but only if Italy granted the same rights (as it did not) to Swiss and French citizens. While most Italians before 1900 had traveled the world without so much as a passport, migration after 1914 involved lengthy interactions with state bureaucracies, at home and abroad.

Migrations from Italy declined between the two world wars but whether those declines were a direct response to restrictions like these is not at all obvious (see Table 6.1). The most obvious and effective damper on migrations from Italy were the two world wars. By contrast, during the 1920s — the main era of new restriction worldwide — the numbers of migrants dropped rather irregularly. At 200,000 to 400,000 a year until 1928, Italy's migrations were still as large of those of the 1880s and 1890s.

In part, this was because three important receiving nations — France, Argentina, and Brazil — hesitated to impose harsh restrictions during the 1920s. Reeling from wartime losses, and fearing the military consequences of a stagnating native population with a relatively low birthrate, France continued to welcome migrants throughout the decade. Italian migration to France actually increased in the 1920s, and Italians soon became the largest group of foreigners living in France.[11] While Argentina continued to worry over the "Italianization" of its large cities, as it had since 1900, it too hesitated to restrict migration. Only the Great Depression pushed Argentina and Brazil temporarily into the restrictionist camp. After 1930, immigration restriction and economic protectionism became the rule everywhere as nation states sought to save troubled industries and to prevent escalating unemployment. Tariff barriers and immigration restriction were twinned responses to global

Table 6.1 Migrations from Italy, 1913–45

Year	*Number of migrants*
1913	872,598
1914	479,152
1915	146,019
1916	142,364
1917	46,496
1918	28,311
1919	253,224
1920	614,611
1921	201,291
1922	281,270
1923	389,957
1924	364,614
1925	280,081
1926	262,396
1927	218,934
1928	140,851
1929	174,802
1930	236,438
1931	165,860
1932	83,348
1933	83,064
1934	68,461
1935	57,408
1936	41,710
1937	59,945
1938	62,548
1939	29,489
1940	51,817
1941	8809
1942	8246
1943	0
1944	0
1945	0

Source: Gianfausto Rosoli (ed.), *Un secolo di emigrazione italiana, 1876–1976* (Rome: Centro Studi Emigrazione, 1976).

economic crisis.[12] Ultimately, it was worldwide depression and the corresponding collapse of the global labor market that ended the mass migrations. By 1933, only France kept its doors relatively open to migrants. And even there, work opportunities diminished as unemployment rose. The numbers leaving Italy plummeted, and reached zero after the onset of World War II.

Table 6.2 Destinations of Italy's migrants, 1915–45 (%)

	*Labor migrants**	*Elite*†	*Exiles*‡
Europe	52	39	92
North America	25	22	3
South America	19	26	5
Africa	3	7	–
Asia	–	6	–
Australia	1	–	–
(Number	4,472,347	188	620)

Sources:
* Rosoli, *Un secolo di emigrazione italiana, 1876–1976.*
† Ugo E. Imperatori, *Dizionario di italiani all'estero (dal secolo XIII sino ad oggi)* (Genoa: L'Emigrante, 1956).
‡ Franco Andreucci and Tommaso Detti (eds), *Il movimento operaio italiano, Dizionario biografico 1853–1943* (Rome: Riuniti, 1975–8).

Restriction completely transformed the context in which millions of international family economies dependent on male labor functioned. Yet, surprisingly, the destinations of labor migrations did not change as dramatically as might have been expected (see Table 6.2). At most, laborers after 1914 were somewhat more likely to seek work in Europe than before the war. While growing in importance, France had long been an important destination for Italians. Between 1876 and 1914, 1.5 million or 12 percent of all migrants had declared their intention to migrate there. In the interwar years, over one-third of the smaller numbers of Italy's newest migrants went to France, and it replaced the U.S. as the most important single destination for Italians seeking work or a new life abroad.

Restriction influenced labor migrations in other significant ways, however. The rising representation of women among migrants — to an historic high of 33 percent — suggests that more families were willing to relocate their home bases abroad as American restrictions increased, making circulatory male migration more difficult.[13] Soaring rates of return, despite Italy's uninspiring economic and political climate, also suggest that reunions in Italy increased. Almost two-thirds of Italy's migrants to Europe returned home in 1921–45 (a number that probably equaled prewar levels).[14] The highest rates of return, however, were from the United States, which (after immigration restriction) turned to Mexican, Puerto Rican, and native-born southern African-American laborers to replace

European workers.[15] Between 1921 and 1945, returners represented fully 83 percent of those who left Italy for the U.S. and Canada. Return rates from Latin America, which had once equaled those from the U.S., were now less than half the North American rate, having declined slightly since the prewar years.[16] Both quantitatively and qualitatively, the international mass migrations had ended.

The Italian state and Italy's diasporas

Italy's government, too, focused increasing attention on its migratory population in the twentieth century, culminating in both Mussolini's hostility to emigration and his desire to transform Italy's diasporas into a ready-made empire. The roots of Mussolini's policies were deep. Throughout most of the period of the mass migrations, the Italian state had asserted an abstract interest in "making Italians" at home and abroad, and in maintaining migrants' ties to the homeland. Before 1890, however, it did little more than count migrants, and oversee their comings and goings. In 1876 the Ministry of Industry and Agriculture began collecting and publishing yearly statistics on migration. Its task was complicated by the fact that Italy required no passport of its emigrants.

By century's end, state activism had increased, but only modestly, and only in the face of complaints and independent initiatives from its two major (and competing) critics — the Socialist Party and the Catholic church. At least since the 1870s, Italy's consuls had been charged with maintaining contact with migrants living abroad. In 1887, Italy's government suggested that its consuls take a more active role among migrants by encouraging diaspora celebrations of the national holiday of unification (Venti Settembre). At the same time, it provided limited funding for Italian language schools abroad. Responding to international criticism of labor recruiter abuses a year later, the Italian legislature also passed a first comprehensive emigration law. This confirmed the essential privacy of a citizen's decision to emigrate but also established the state's interest in protecting its citizens — a task that had been largely left in the past to the consuls. The law provided for regulation of ticket agents and labor contractors and it prescribed health standards for transatlantic steam ships. A second emigration law passed in 1901 provided for expanded governmental protection of migrants, allowing the temporary suspension of migration to Brazil in 1902 and to Argentina in 1911 and 1912. The same law created a General

Commission on Emigration that collected and published information about and for migrants in its own extensive bulletin.[17]

The 1901 law on emigration was one of the first Italian pieces of protective welfare legislation and an early sign that Giolitti's liberal government would move beyond the laissez-faire assumptions of the founding generation of Italian nationalists to respond to its Catholic and socialist critics. (See also pp. 114 and 126–7 in Chapter 5.) Before Giolitti, protection of migrants had been left to non-governmental organizations called patronati, many of them sponsored by Catholics or socialists hostile to the state. Over the next twenty years, both groups would begin a long process of accommodation with Italy's state, and seek funding for their work with migrants, generating new forms of nationalism, less attached to the Risorgimento and to civiltà italiana, in the process.

Of the patronati working with emigrants, both lay Catholic and Vatican-initiated projects developed first. In 1875 a congress of Italian economists meeting in Milan created a lay Catholic organization, modeled on Germany's St. Raphael Society, to provide aid to Italian migrants. It established contacts with migrants' mutual benefit societies and with natives' charitable organizations in Canada, the U.S., and Argentina, and it published a bulletin for potential migrants at a time when Italy's government (see Chapter 2) had scarcely begun to come to terms with the scale or causes of emigration.[18]

In 1887, on the urging of the Vatican, Giovanni Battista Scalabrini, the Catholic Bishop of the northern city of Piacenza, founded the Pious Society of Missionaries of St. Charles Borromeo (the Scalabrinians) to work with migrants to the Americas. Scalabrini was dismayed that the Italian state did so little to protect migrants. Like other clerics of his era he also distrusted any activism undertaken by a state that had separated itself from the church and from religious principles. Pope Leo XIII had earlier noted American Catholic bishops' limited missionary work among Italians, Negroes, and Indians, and in 1888 he issued an encyclical letter to the American bishops informing them of Scalabrini's initiative. Opening a mother house in Piacenza and naming it after Christopher Columbus, Scalabrini sent his first missionaries to Brazil and to the United States in 1888. The order then established over 100 American missions, including several agricultural colonies in Brazil and the United States.[19]

The Scalabrinian missionaries represented in the other Italies the rise of social Catholicism, associated in Europe with Leo XIII's

Rerum novarum encyclical of 1891. Increasingly, clerics in Italy became involved in activities on behalf of Italy's poor and laboring classes. After 1905, the Vatican gradually abandoned its intense hostility to Italy's state, allowing Catholics to vote, to hold political office, and even to organize political parties. Much of Catholics' political activism in the twentieth century developed in opposition to socialist and liberal initiatives, culminating in the formation of the People's Party ("popolari") before World War I. Foreshadowing this activist and populist Catholic nationalism was the flag Scalabrinians adopted in the 1880s. It promised to strengthen faith and homeland — "religione e patria" — simultaneously.

Like Scalabrini, who died shortly after the turn of the century, his friend and collaborator Geremia Bonomelli, the Bishop of Cremona, and the founder of the Society for Assistance to Italian Workers in Europe and the Levant (the Opera Bonomelli), believed that religiosity and love of homeland complemented each other. Bonomelli's Opera, centered in Milan, joined clerics and lay supporters like the Egyptologist and nationalist Ernesto Schiapparelli. Concerned with the growing influence of socialists, especially in northern Italy, Bonomelli established his Opera to compete with Milan's socialist Società Umanitaria (see Chapter 5) and to spread Catholic civilization among the migrants internationally. Opera Bonomelli, in turn, pushed Umanitaria to work with migrants in 1902. Both groups sought to aid migrants while at the same time incorporating them into the nation of Italians as either socialists or Catholics. The two groups competed quite directly in Europe to provide advice, aid in railroad stations, and social services. After several years of furious polemics and competition, both eventually received government funding for their work, a sign that each was overcoming its separate tradition of hostility to the Italian state, and developing its own accommodation to a national Italian identity attached to a national state.[20]

In 1907, Italy's General Commissioner of Emigration began his own initiative among migrants and called a first world conference on Italians abroad in Naples (followed by another after World War I). At the 1907 conference, all participants were residents of Italy, mainly men active in the patronati and in the government agencies working with migrants. Seeking to encourage migrant participation, Schiapparelli in 1909 went on to found Italica Gens, an organization for Italians living abroad. A secular organization, Italica Gens attempted to work through Schiapparelli's connections with cleric missionaries in the Americas. The organization published its own magazine, of the same name, and had offices in 157 cities in North America, South America, Europe, and Africa. Its purpose was to cultivate love of the homeland and to promote civiltà italiana, a goal it shared with the slightly older patriotic and secular society, the Società Dante Alighieri. Founded in 1895 by Italian nationalists in multi-ethnic (and still Austrian) Trieste, the Dante Alighieri linked a new generation of secular Italian nationalists to the civilizing mission of civiltà italiana. It too opened offices in centers of major Italian settlement, offering Italian language and culture programs.[21]

Unlike most Catholic and socialist activists of the 1880s and 1890s, the new nationalists of early twentieth-century Italy often dreamed of the other Italies as colonies in a new Italian empire. The idea was not completely new. Leone Carpi, the Risorgimento activist and exile, for example, had titled his 1870s review of consular reports on Italians abroad *Delle colonie e dell'emigrazione d'italiani all'estero.*[22] In 1898 the prominent liberal economist Luigi Einaudi developed his own interpretation of Italy's colonies in *Il principe mercante: Studio sull'espansione coloniale italiana* ("The merchant prince: a study of Italian colonial expansion") which described the petty traders of the Argentine Plata as the equivalents of medieval Venetians.[23] In the same decade, Francesco Crispi — Italy's earliest proponent of colonial expansion — concurred. Emigrants, he wrote, "must be like arms which a country extends far away into foreign places to draw them into its orbit of labor and exchange relations; they must be like an enlargement of the boundaries of its actions and its economic power."[24]

Italy had repeatedly but unsuccessfully tried to build a more conventional empire for itself in nearby Africa in the late nineteenth century. Support for colonization in Africa — for example in Tripoli in 1883 — was especially strong among southern landowners who saw it as an alternative to underemployment and emigration for southern peasants. Southern Italians migrating voluntarily to Tunisia outnumbered French settlers there in 1880, causing one wag to term it "an Italian colony ruled by France." In 1889, Italy also became the protector of Somalia, and a year later it constituted Eritrea as its first colony. The revolt of the Eritreans, who defeated Italian forces at Adua in 1896, quickly ended this first experiment in Italian colonialism, forcing Prime Minister Crispi from office and humiliating a generation of hopeful imperialists seeking to build the Italian nation through territorial expansion.[25]

Italy's newer nationalists — lay Catholics, some clerics, even some socialists — soon became enthusiastic supporters of a new round of

Italian empire-building. In 1910, a congress of Italian nationalists heard the outspoken critic of liberal Italy, Enrico Corradini, call emigration a phenomenon "if not of an inferior people, at least of a people at an inferior stage of existence."[26] Nationalists agreed with Luigi Villari that Italy's empire would not be "mere greedy imperialism . . . but an alternative to unemployment and starvation," especially for southerners.[27] Imperialism, Corradini concluded, was a necessary development for a "proletarian nation" trying to escape global subordination from "the French, the English, the Germans, the North and South Americans . . . who are our bourgeoisie."[28] Under pressure from such nationalists, Italy's government funded a Colonial Institute in 1906. And in 1911, Italy's war with Turkey gave the country yet another colony — this one in Libya, the "fourth shore" (nationalists claimed) of an emerging new Roman Empire.[29]

This growing tide of nationalism in turn encouraged Italy's state to clarify its relationship to its migrated citizens. Responding to complaints heard at the first migration conference, Italy's legislature in 1912 carefully defined the parameters of Italian citizenship, and it recognized those born in Italy but living abroad as citizens. Italy even accepted the dual citizenship of those who had chosen to become citizens of the nations where they worked, and it accepted children born abroad as Italian, too. The law thus established that descent and blood (*jus sanguinis*), not residence or birthplace (*jus soli*), defined the Italian nation. It put aside Mazzinian and Garibaldian notions of a voluntary nation, based on love of patria (a place), civiltà italiana, or military service, to define the Italian nation as a biological group, or stirpe (race or tribe). Since Italy's legislature granted suffrage to men over thirty at the same time (universal manhood suffrage would follow in 1919), it was careful also to note that male citizens could not vote so long as they lived abroad.[30] Socialists in the P.S.I. and Catholics in the People's Party (the largest of the new mass parties) quickly moved to recruit returning migrants but neither had any incentive to organize in the other Italies.

In a related, if quiet, move, the Italian state also moved to substitute its own efforts for those of thousands of migrant padroni directing Italy's labor migrations to countries needing their labor. In 1904 Italy negotiated the first of a series of bilateral treaties governing migration and commerce, first with France, and later with Brazil, Holland, Germany, and other countries. Exact provisions varied, but most treaties specified the number of migrants to be admitted and offered some guarantees of protection (wage levels, safety from discrimination, and welfare provisions) for migrants.[31] Just after World War I, Italy's last liberal government also proposed that the International Labor Organization set standards for migrant workers' access to welfare programs worldwide. Repeatedly, however, receiving nations rejected international regulation of these matters, which they viewed as attacks on their sovereignty, and their ability to define citizenship rights as they pleased.[32]

Almost alone among the former internationalists of the world, the Vatican, along with most Italian socialists, remained neutral during World War I. Still, Italy's belated entrance into the war (in the hopes of "redeeming" Italian-speaking districts still under Austrian rule) also served as a sharp reminder of the growing importance of national sentiments in a country once known largely for their absence. Other Italies had coalesced abroad by the second decade of the century, too. As Italians debated the fate of Fiume in the aftermath of the Versailles Treaty, followed with some interest the escapades of nationalist Gabriele D'Annunzio in the Adriatic, and debated the threat posed by workers' councils seizing northern Italian factories, the other Italies seemed of little import, even to nationalists. The rise of Mussolini, however, would bring them close to the center of interwar debates about the Italian nation.

Forging a fascist diaspora

The rewards of creating a fascist diaspora were surely appealing to Benito Mussolini as an ambitious nationalist leader of a relatively small country. The number of Italians living abroad had peaked by 1920; the nine million Italians living outside the country raised Italy's total population by 25 percent. The children of the migrants raised it still higher. Mussolini's policies toward the diaspora began with his efforts to curtail emigration; they continued with his determination to channel Italy's underemployed toward fascist colonies in Africa. The intent of fascist state policy toward Italians already living abroad also quickly became clear. Hostile to most new emigration, Mussolini's government remained deeply interested in the other Italies. It saw them, too, as colonies — but with the wealth to support Italy's expansion in Africa.

A vociferous interventionist and self-proclaimed nationalist by the end of World War I, returned migrant Benito Mussolini waited several years after seizing power (in 1922) before tackling the complex issue of migration. Even as he arranged the murder of political enemies like the socialist Giacomo Matteotti and destroyed the

Italian legislature, his government in 1924 sponsored an international conference on the problems of immigration and emigration.[33] In 1927, Mussolini suspended the General Commission on Emigration as too independent to conform to his own plans for Italy's foreign policy. By 1931, he had created a new agency — the Commission for Migration and Internal Colonization — located within his Foreign Ministry in order to guarantee that migration served the broader purposes of Italian imperialism.[34] Although he opposed migration generally, Mussolini did not hesitate to negotiate bilateral treaties to facilitate it when he believed migration served the nation's interest. One treaty sent Italian laborers to Germany.[35]

While Mussolini's policies had but little impact on labor migrants whose decisions and migration patterns instead reflected restrictions in the receiving countries, they did influence the migrations of both political dissidents and elite Italians in the 1920s and 1930s. Europe continued to attract the overwhelming majority of political dissidents — a familiar pattern at least since the Risorgimento. Emigration rates among Italian labor leaders soared once Mussolini came to power — 80 percent (compared to only 30 percent in the years between 1870 and 1914) left the country for at least a few years. More than 300 of Italy's most prominent labor leaders and anti-fascist political activists lived abroad for most of the interwar years. Fully one-third of Italy's exiles (fuoriusciti) went to France. Thus, under fascism, workers and exiles alike could claim "we're going to France to find work and liberty."[36] The Soviet Union — which held no attractions whatsoever for labor migrants — drew 17 percent of the exiles, mainly members of the Italian Communist Party (P.C.I., founded in 1921). Tiny Belgium and nearby Switzerland each drew more anti-fascist labor leaders than the United States, Argentina, and Brazil combined.[37] Catholic anti-fascists — an admittedly small, but nevertheless influential group — went to England and the United States more often than did their left-wing and secular counterparts.

In a curious way, Italy's elite migrants seemed to follow the paths Mussolini wished for them. In striking contrast to other migrants, Italy's elite found Europe even less attractive than they had before 1914. Italy's universities educated twice as many students under fascism, yet jobs were not always available for graduates. Unlike the exiles and labor migrants, elite migrants looked for opportunity in the developing world. Only 20 percent went to France, while the proportions going to Asia, Africa, and Latin America increased once Mussolini came to power. Dreams of influence in less developed countries were scarcely new to Italy's migratory elite, given the legacy of civiltà italiana. But more than Italy's past rulers, Mussolini explicitly linked them to his imperial dreams of dominance over inferior peoples. Mussolini consciously used state power to build a national diaspora that generations of free migration had failed to produce.

After coming to power, Mussolini built on earlier precedents to consolidate and expand his influence among migrants. Since voting soon disappeared as a right of citizenship, the Fascist Party was more interested than the prewar liberal parties in organizing abroad. In 1923 the Fascist Party moved to encourage the organization of fascist clubs abroad. And in 1924 the government closed down the older welfare societies, Umanitaria and Bonomelli, claiming that fascist consuls could better protect, advise, and serve Italian citizens abroad.[38] Mussolini's pursuit of new colonies in Africa generated propaganda publications with titles like *L'Italia coloniale* (1924–43), *L'Italia d'oltremare* (1936–43), and *L'Italia e il mondo* (1927–8). All trumpeted the important leadership role Italians could play in Africa. In an effort to encourage migrants who might otherwise go abroad to remain in the countryside, Mussolini also funded a few new agricultural colonies in rural Piedmont, the Agro Romano, Sardinia, and Sicily.[39]

Appointing his supporters to consular posts, Mussolini urged them to expand their activities among the migrants, especially through cultural programming that emphasized the unity of Italians everywhere. Consuls expanded their financial support for traditional programs like Italian-language schools, Italian-language newspapers, and the Dante Alighieri Society. Newer initiatives included the formation of groups like the Opera Nazionale del Dopolavoro all'Estero (which promoted leisure-time activities, as did the popular mass organization in Italy) and the Fondazione Nazionale Figli del Littorio for children. In several countries, consuls worked actively with local supporters to raise funds for a Casa d'Italia (Italian House) which one critic described as "a little patria fascistically integrated into the Italian emigrant community."[40] Back in Italy, fascists promoted national loyalties by funding summer camps for children born to Italians abroad. By the end of the 1930s, there were 487 local fasci worldwide, backed up by a multitude of periodicals, as well as scores of pro-fascist charity, religious, athletic, educational, and veterans' organizations. Fasci abroad after 1928 claimed women's sections and separate organizations (Balillas, modeled on the fascist youth organization in Italy) for boys.[41]

Just as the nationalism of the Risorgimento had developed along diverging paths at home and abroad, so too Mussolini's efforts to create diaspora nationalism had diverse consequences in Italy's many diasporas. At home, Mussolini hoped to eliminate the still visible regionalism of Italy's peoples. Abroad, he faced an even wider range of enemies. Natives of the countries where Italians lived and worked usually objected to fascist political organizations on their soil. Labor migrants also heard very differing messages in Mussolini's bombastic statements about the glories of his new Italian Empire. And almost everywhere as fascist consuls worked to spread Mussolini's message to migrants, they found themselves shouted down by the angry voices of anti-fascist exiles driven from Italy into the diaspora.

Fascism, anti-fascism, and diaspora nationalism

Three cases — the U.S., France, and Africa — portray the variety of battles between fascists and anti-fascists over the meanings of diaspora nationalism in the interwar years, and the differing outcomes of their conflicts. Africa best represents the labor migrants' and their children's resistance to fascist nationalism, while the United States symbolizes Mussolini's best successes. France, by contrast, shows how anti-fascists and labor activists became important leaders of Italians in this country, creating an alternative, and democratic, diaspora nationalism that influenced events in Italy after Mussolini's demise. The consequences of migration patterns on these differing outcomes were probably significant. In France, fuoriusciti exiles and labor migrants mingled in impressive numbers. Elite Italians, inspired by Mussolini's empire-building, established a good presence in Africa, but encountered few labor migrants, and even fewer anti-fascists, there. In the U.S., American hostility to labor radicalism of any variety gave elite migrants and fascists, joined by Catholic parish priests, a subtle advantage in their own battles with anti-fascist labor leaders.

Had Mussolini read the statistics so carefully gathered by the Italian government since the 1870s, he might have guessed that few labor migrants would be impressed by slogans like "today, the real America for all Italians is Libya."[42] Despite enthusiasm among Sicilian landowners for colonization in the 1890s, only 4000 Sicilian peasants had settled in Eritrea by 1900. By 1940, the numbers had risen only to 70,500. Another 110,000 lived in nearby Libya. Collectively, Italy's colonies had attracted only

305,000 settlers by 1940 — less than the number of Italians living in either New York or Buenos Aires, and only slightly more than the number of migrants who left Italy in any year during the 1920s.[43] Mussolini's foundation of an agricultural colony in Tripoli in 1926, and his attack on Ethiopia in 1935, did little to make these colonies more attractive to workers, especially once the Ethiopians began their rebellion against Italian occupation.[44]

Elite migrants seemed far more interested in Italy's empire than labor migrants, perhaps because professional opportunities to work in Italy stagnated under fascism. In any case, the backgrounds of elite migrants changed significantly during Mussolini's rule. Businessmen replaced artists as the largest group (soaring from 15 to 26 percent of the total). In second place, visual artists (16 percent) displaced performing artists. Scientific and technical workers were even better represented (15 percent) than before the war, while missionaries and church officials held steady at 6 percent. Business, technical, and scientific workers undertook projects in many economically less advanced corners of the world — not just in Africa and Italy's colonies there, but in Asia and Latin America too. These were a diverse group of men, not all of them fascists themselves. They included missionary Giuseppe Taufer (who worked in a leper colony in Rhodesia), architect Carmine Trotta (who designed a new garden city in North Africa), and Romolo Cipressi who opened a munitions factory in Yemen in 1927.[45]

If fascism drew mass support anywhere among Italian migrants and their children, it did so in the United States. In fact, the first fascio in New York had formed among Italian war veterans and community activists (including some in the Sons of Italy) even before the first visit to America of fascist legislator and organizer Giuseppe Bottai. The organization of fascists in New York in 1921 thus predated Mussolini's March on Rome the following year. Once in power, Mussolini feared that diaspora fascists were too independent of his control, and he sought to subordinate them within the centralized Fascist Party in Italy. Predictably, the incorporation of an immigrant organization into a foreign political party drew immediate criticism from U.S. officials, and a New York headline trumpeted "Fascists invading the U.S. as part of world-wide expansion." In 1925 the original Fascist Central Council for the U.S. changed its name to the Fascist League of North America and claimed complete autonomy from Italy's party. Subject to continued harsh criticism, however, it too disbanded in 1929. At that date, fascists in the U.S. probably numbered only about 3000–5000, and they were under attack both from natives hostile to foreign interference and from a vocal anti-fascist movement.[46]

By the late 1930s, anti-fascist exile and Harvard Professor Gaetano Salvemini estimated that there were twice as many open anti-fascists — 10 percent of the Italian population in the U.S. — as there were open fascists.[47] Recent exiles like Salvemini were not particularly prominent as anti-fascist spokesmen in the U.S.; after all, only 8 percent of the leading fuoriusciti from Italy went to the Americas. New exiles — most of them socialists or liberals associated with the group Giustizia e Libertà (see below) — added their voices to the anti-fascist cause spearheaded by earlier labor migrants. Radical exiles did not always fare well in the U.S. in the 1920s. Some — like labor organizer Giovanni Pippan — were mysteriously murdered while others, including Vincenzo Vacirca and Ezio Taddei, were expelled by the American authorities for their radicalism. Instead, the tiny minority of Italy's exiles who were moderate socialists (like Giuseppe Modigliani and Emilio Lusso), liberals (like Salvemini), and anti-fascist Catholics like Don Sturzo found refuge, and freedom of speech, as anti-fascist activists in the U.S.

Prewar migrants, like anarchist and syndicalist Carlo Tresca and communist Vittorio Vidali, were more prominent and numerous in the campaign against Mussolini, and most had pointed to the dangers of fascism in Italy already in the early 1920s.[48] However, by the mid-1920s, the most important anti-fascists in the U.S. were modest Italian and Italian-American activists in the American labor movement. The largest group was in the two garment worker unions with large Italian-American memberships, the I.L.G.W.U. and the A.C.W.A. Union activists like Luigi Antonini (from New York) and Tampa activist Giovanni Vaccaro quickly founded the Antifascist Alliance of North America in 1923.[49] The anti-fascist movement in the U.S. enjoyed popular support, too: Italian police maintained dossiers on about 6000 subversives living in the U.S.; the group included both long-time residents and recent arrivals from Italy. Most were young men. American anti-fascists remained in contact with anti-fascists at home and in France, and eventually participated in a number of transnational political actions, notably the fight against fascism in Spain.[50]

Despite the activism and numerical superiority of the articulate anti-fascists, Salvemini believed that the smaller group of fascist apologists enjoyed greater popular support among Italian immigrants. In the 1930s, he estimated that over a third of the 4.6 million Italians in the U.S. were potential fascist recruits. By this he meant that they willingly listened to endless positive messages about Mussolini from what Salvemini called the "fascist transmission belts."[51] Beginning with the Sons of Italy, many mutual aid societies and their leaders openly praised fascism and welcomed the overtures of Mussolini's consuls. Salvemini regarded Generoso Pope and his newspaper *Il progresso*, along with most of the prominenti of Italian-American settlements and activists in the Sons of Italy, as fascist sympathizers. Surprisingly, their sympathies with Mussolini drew as much on their American experiences and Americanization as on their roots in Italy.[52]

Unlike the anti-fascist immigrant socialists and labor activists, Mussolini (like many prominent Americans in the 1920s) was an outspoken opponent of bolshevism and communism. Italians in the U.S. had watched the execution and deportation of too many Italian-born radicals to misunderstand Americans' intense hostility to labor radicalism. Even more important, Italian immigrants in the U.S. (and Canada and Australia, too) seemed to see in Mussolini's Italy a homeland of which they could be proud. However bombastic, Mussolini provided a symbolic counter to American prejudices about backward and inferior Italians unworthy to enter the country. Smarting from the stigma of restriction, migrants saw in Mussolini someone who proudly defended their worth. Mussolini certainly encouraged the nationalism of migrants in the U.S. when he intervened in defense of Sacco and Vanzetti — anarchist radicals he would himself have imprisoned or murdered at home.[53]

Aware of American distrust — but also of Americans' begrudging fascination with the figure of Mussolini as the most successful anti-bolshevist of the 1920s[54] — fascist consuls regularly argued that Italian migrants could and should become good American citizens. As citizens, they could use their influence to build sympathy for the new, and anti-socialist, Italy and its fascist ruler. Rudolph Vecoli has recently linked the "fascistization" of Italian-Americans with their Americanization during the interwar years. Two dimensions of Italian nationalism — a patriotic love of the homeland and an ethnic but firmly American identity — proved mutually reinforcing.[55]

Still, the most important "transmission belt" for fascist ideas in the U.S. was probably the Catholic church with its national Italian-American parishes and parochial schools. Priests in Italian-American communities played an especially important role in fostering sympathy for fascism once the decades-long feud of Vatican and Italian state ended. Attending churches organized by Italian missionaries (rather than subservient to the Irish- and German-origin bishops of the American Catholic church), Catholic Italians in the U.S. repeatedly heard positive messages about fascism from their priests. Reconciliation with the fascist state was a central issue in the Italian-speaking Catholic world of the 1920s as Mussolini and the Pope negotiated the Lateran Accords that made Catholicism the state religion of fascist Italy. Parish schools used fascist schoolbooks in their Italian classes, and these taught a history of civiltà italiana that culminated in the new fascist empire. Peter D'Agostino, historian of immigrant Italian Catholicism in the U.S., has concluded that the church, too, celebrated "italianità, as they gave devotional sanction to Fascism as a form of 'Americanism.'"[56] Again, fascism and new American national identities seemed perfectly compatible.

By 1935, when Mussolini attacked Ethiopia, Italian-Americans seemed wild with enthusiasm for their newly powerful homeland. They collected more than $500,000 to aid the Italian effort. Anti-fascist efforts to subdue immigrant support for imperialism foundered on American fears of communism and radicalism. The anti-fascist Italians of New York desperately sought to build a multi-ethnic alliance with African-American leftists to counter fascism's growing hegemony in New York's East Harlem, where the two groups lived as near neighbors.[57] Neither group found much popular support. Within Italian- and African-American communities in New York, diaspora nationalism had become racialized (in the African-American case, notably among supporters of Marcus Garvey).[58] Dividing by race, in defense of far-away ethnic homelands, rather than uniting in a multi-ethnic left-wing alliance, both groups demonstrated just how American they were.

Even when the U.S. declared war on Italy in 1941, anti-fascist Italians were shocked that the U.S. was little concerned about the loyalties of openly fascist leaders of Italian communities. The U.S. interned a few Italian migrants, as did Australia and Canada. They were not — as were the Japanese — imprisoned collectively, however, on the basis of their race, or — as the anti-fascists wished — on the individual basis of their political loyalties.[59] Instead, Italian citizens were required merely to register, or, as citizens of an enemy country, to report to the government as "enemy aliens." Those who had naturalized and acquired U.S. citizenship escaped all surveillance. Former apologists for Mussolini like Generoso Pope, the newspaper editor, not only remained free but became actively involved in American campaigns to guarantee that fascism's defeat in Italy did not open the doors to a left-wing revolution. He did this, furthermore, as the old syndicalist revolutionary, Carlo Tresca, was gunned down on the streets of New York — the last, not the first, act of violence sparked by conflicts between Mussolini's supporters and opponents in Italian America.[60]

In France, by contrast, anti-fascist exiles achieved what Italian-American opponents of Mussolini's campaign could not do in East Harlem. They gained popular support from Italian migrants, making France — and particularly Paris — the capital of a transnational anti-fascist movement. And in France, Italian anti-fascists also found support and fellowship with French labor activists, especially during the period of the Popular Front (1934–9). In France, the anti-fascist movement was not only transnational; it was also multi-ethnic. Just as Italy's socialist exiles in the 1890s had found common cause with European socialist labor movements, exiles from fascist Italy — the largest group of them communists and socialists — found workers in France and Belgium eager to embrace anti-fascism.[61]

As Mussolini's black-shirted squads attacked labor institutions throughout Italy in 1921, exiles in France attempted to counter the spread of fascist propaganda by providing an alternative definition of italianità, based on an international struggle against exploitation and oppression.[62] Already in 1923 former Prime Minister Francesco Nitti arrived to alert French readers in Switzerland and then Paris about the terrors of Mussolini. With the 1924 murder of socialist deputy Matteotti, French and Italian political activists alike recognized that Mussolini would brook no opposition from democratically elected officials. Self-consciously republican, French citizens became alarmed at the rise of this new dictatorship to their south. The French were not completely immune to the kinds of fears Americans also demonstrated toward "foreign" politics on their soil. By 1926 a variety of anti-fascist journals in Italian circulated in France (which closed down the *Corriere degli italiani* after it recommended Mussolini's assassination). But anti-fascism quickly became a popular cause of the French generally, opening a sympathetic space for Italian exiles of a wide variety of ideological persuasions to pursue their own goal of defeating fascism at home.[63]

During the mid- to late 1920s, the grandsons of Garibaldi promised to raise a liberation army in France. Radical democrats, democratic reformers, and a few anarchists in Paris — Alessandro Bocconi, Alceste DeAmbris, and Luigi Campolonghi (an exile of the 1890s) — joined together in the Italian League of the Rights of Man (modeled on a French organization). Its leaders then formed the non-communist but also anti-fascist Concentrazione Antifascista (C.A.) that also attracted socialist exiles like Bruno Buozzi, Marco Riccardi, Claudio Treves, and Alberto Mesci. (The exiled communists of Italy — still committed to building resistance within Italy and working clandestinely there — organized independently, more often in Moscow.) In 1928, exile Carlo Rosselli launched his own "supraparty" movement (again for non-communists) called Giustizia e Libertà (G.L.). Collectively, the anti-communist G.L. and C.A. and the communists of the P.C.I. formed three wings of an Italian- and exile-dominated diaspora resistance in France and the Soviet Union. Their members were all Italian. Recent exiles dominated their discussions, without significant support from ordinary migrants working in France, and without much active cooperation from French activists.

By the mid-1930s, however, Italian workers (who accounted for over half of the 400,000 foreign members of the French union C.G.T.) and the formation of a Popular Front brought together French and Italians in a multi-ethnic anti-fascist movement. French and Italian communists figured prominently in all these anti-fascist initiatives.[64] Fascist enthusiasm in France developed only among the far smaller group of Italian professionals, shopkeepers, small businessmen, employees of Italian companies, the diplomatic corps, and Catholic missions — not among workers.[65] French left-wing organizations and unions saw in anti-fascism a rallying cause for all workers, and both groups sent large numbers to fight in Spain, both as part of the all-Italian Garibaldi Brigade, and in the International Brigades.[66] For the "brigadisti," the unity of internationalist and nationalist struggles seemed obvious. They fought, they claimed, "Today in Spain, tomorrow in Italy."[67]

Especially after the shocking murder of the exiled Rosselli brothers in 1937, Italian workers in France joined anti-fascist Popular Front initiatives in large numbers. In 1937 the main currents of the exiled anti-fascist left in France and Belgium came together to found the Unione Popolare Italiana (U.P.I.). The U.P.I. gave institutional expression to immigrants' commitment to help defend the countries they shared with native workers. In the spring of 1939, at the U.P.I.'s Lille convention, 700 delegates urged Italian migrants to join the French army to fight fascist expansionism, after which they adjourned by singing the French national anthem, the "Marseillaise."[68] Italian workers displayed their identification with a labor-dominated and multi-ethnic anti-fascist movement by learning to sing the "Internationale" in French. Meanwhile, French and Belgian workers also made the Italian revolutionary song, "Bandiera rossa" ("Red Flag") part of their country's left-wing cultural repertoire. It was in France, not in the U.S.S.R., that exiled Italian communists explored some of the cultural initiatives that would mark them as distinctive (later called "Euro-") communists in the postwar era. Significantly, they first created the folk festival of l'Unità (honoring their newspaper) in France during the Popular Front years.[69]

Elsewhere around the world, the conflicting claims of fascists and anti-fascists, and diaspora nationalism, fell between the French and American extremes. Mussolini enjoyed far more support in Australia and in Canada than in the Latin world.[70] The middle-class Italian migrants of Brazil showed some sympathy for Mussolini, and their children were also attracted to Brazil's local fascist party. Rather than support an active anti-fascist movement, however, most working-class Italians in Brazil merely remained aloof from the Italian consul's demands for loyalty and financial support.[71]

Anti-fascist initiatives were more common in Argentina, but the influence of fascism has received too little attention from scholars there to allow firm conclusions.[72] Argentina's anarcho-syndicalist labor movement had collapsed after the "tragic week" of 1919, provoking a decade of political chaos. By the 1930s, new industrial unions were only beginning to form. In Argentina even more than the U.S., however, Italian anti-fascists claimed the mantle of Garibaldi and Mazzini — figures with great appeal to the traditionally republican leadership of Italian institutions there.

Worldwide, both fascism and anti-fascism became competing and transnational but self-consciously nationalist movements seeking to bind migrant Italians to Italy. They advanced significantly different models of Italian nationalism. Fascism argued that blood and racial pride linked Italian migrants to an anti-democratic and imperialist national state that claimed, but also deeply transformed, the centuries-old civilizing mission of civiltà italiana and the Catholic church. Anti-fascists, too, sometimes referred to the glories of the Italian past, but they preferred the ideals of the Renaissance and of humanism to those of Rome and the church. Whether republicans and liberals or socialists and communists, anti-fascist exiles called upon migrants to join with other "real Italians" to restore democracy and justice and to end oppression in Italy and worldwide.

Even internationalists, and communist activists in the Comintern, had few reasons for criticizing the nationalism inherent in diaspora anti-fascism. For them, anti-fascism was a twentieth-century Italian expression of national liberation — a goal that the Comintern — unlike the Second International — had accepted as compatible with proletarian revolution worldwide after considerable theoretical debate in the years after World War I. Worldwide, leftists had again forged a new accommodation of international and national aims. Support for revolutions in single nations, based on nationalist and anti-imperialist claims, had displaced Marx's expectation of a global proletarian revolution against the power of capital.

The linked histories of fascism and anti-fascism as global movements reveal a striking symmetry. But they also shed light on the diverging experiences of migrants in different receiving countries, with differing national labor movements and with differing political cultures. Whether in France or in the U.S., natives viewed political activism rooted in a foreign country, and undertaken by non-citizens, with some suspicion. Although Mussolini dreamed of politically "fascisticized" colonies, receiving countries generally did what they could to limit his — and his opponents' — political activities among migrants on their territories. Fascism developed its greatest influence in the diaspora as a cultural, not a political, movement.

Organizationally, too, the cultures of the receiving countries shaped the battles between fascist and anti-fascist nationalists. In the U.S., both fascism and anti-fascism remained movements within the other Italies, and anti-fascism seemed as dangerously foreign to natives as fascism. Under these conditions, the diaspora nationalism that exiles and fascist consuls alike drew upon encouraged the development of ethnic and American identities. Indeed this would be the long-term legacy of the political battles of the interwar years in the U.S. In France, by contrast, anti-fascism (and to a much lesser extent fascism) turned into multi-ethnic movements of French, Italians, Germans, Poles, and Spaniards. Anti-fascist activism as easily facilitated migrants' decision to naturalize and to become French citizens as to return home to an Italy where they could again exercise their expanded rights as Italian citizens.

7

POSTWAR ITALY: FROM SENDING TO RECEIVING NATION

> "Mine will be the last generation in this family to work with our hands."[1]

In 1943, the residents of Italy again found themselves invaded and divided politically into three fragments, ruled by Germany, by a much-chastened Mussolini, and by King Vittorio Emanuele III. Those who had tolerated Mussolini's imperialist nationalism had learned a harsh lesson, while Italy's anti-fascists faced the daunting prospect of unifying the "real Italians" to create the Italian nation they had defended through their years in exile. To the surprise of many, the communist leader Palmiro Togliatti, returning from years of exile in Moscow, committed his party and the resistance movement it dominated to a popular front effort to reunite Italy. Ignoring sharp differences among Catholics, communists, socialists, and liberals, anti-fascists determined to create a democratic, but national, state. With fascism prostrate, nationalism still prevailed.

Two years later, Italy was a reunited country but devastated by war, its national economy in ruins. For most Italians, the main challenge was to find enough to eat, not to puzzle over the future of the nation. Exiles flocked home to participate in the new republican experiment. Women became voters. Under the watchful eye of a tense postwar world dividing between supporters of the U.S. and the Soviet Union, Italy's Christian Democrats, the descendants of the "popolari," built a mass party (the D.C.) that governed for forty years. Communists in the P.C.I. also built a new mass party but remained on the margins — powerful, but in perpetual opposition.[2]

Italy changed dramatically in the postwar years. So did Italians, with their long tradition of transnationalism as a way of life. Even Mussolini had conceded that emigration was "a physiological necessity for the Italian people."[3] That pent-up necessity reasserted itself at war's end, and a new era of mass international migrations appeared inevitable. With seven million migrants over thirty years, emigration was impressive, but it fell below the levels of the early twentieth century. The creation of new village-based diasporas faltered, then ceased, and a new, much smaller, but firmly national, diaspora of elite Italians eventually replaced them.

In part, the demise of Italy's prewar diasporas was a consequence of continued state activism. Italy's democratic rulers (like Mussolini before them) intervened in the lives of ordinary citizens to pursue national interests. Pressured by the P.C.I., the D.C. cautiously oversaw the creation of a welfare state that provided some of the same guarantees of economic security available in other countries.[4] However, Mussolini's "warfare state" gave way not only to a postwar welfare state but also to a new padrone state. Well into the 1960s, Italy's government functioned as a labor agent to direct its migrants to foreign work places.

Postwar migrations also reflected fundamental economic change at home. In 1945, few imagined that just fifteen years later the country's economic development would astound the world with a genuine "Italian miracle." Italy's gross national product soared after 1955, dragging popular levels of material well-being along with it. By 1990, Italians boasted that their country was the fifth most important industrial economy in the world. Some claimed Italy's standard of living surpassed that of northern Europe. Once again, the cultural products of Italy became desirable and they could be purchased — as fashion, film, and design — around the world. Few attributed Italy's economic miracle to the Italian state; between 1948 and 1968 Italians saw twenty governments come and go, although their personnel changed only imperceptibly over time. Nevertheless, Italy's economic successes, national systems of mass communication and education, expanded consumer choices, and welfare guarantees did forge stronger, positive connections between state and citizens, fostering a new sense of national solidarity.

Mussolini had told Italians to remain at home to make the nation strong. He was only partially wrong. Almost every Italian moved somewhere during Italy's economic miracle. But unlike the past — when international migrations outstripped moves within Italy — most Italians now found new homes and work in Italy. By far the largest group left the rural South to work in the center and North. After 1970, migration abroad dropped precipitously and permanently, and foreign workers appeared in Rome, Milan, and even Palermo. Italy's transition from sending to receiving nation had taken little more than a decade, and the nation seemed poorly prepared to face the change.

Agrarian reform, the padrone state, and new migrants

Italy's postwar migrations developed in three distinct phases, as the country moved from wartime hunger and postwar hardships toward its economic miracle.[5] In the first ten years after the war, Italy's democratic government did more than construct a rudimentary economic safety net. It turned its attention to agrarian reform and to international diplomacy on issues related to migration. Many of the fundamental characteristics of Italy's proletarian diasporas reasserted themselves during this decade. International family economies survived the transition from padroni to padrone state. But only in the Americas, and among southern Italians, did new village-based diasporas emerge as families relocated home bases abroad.

Italy was scarcely one of the poorest countries in the world in 1945. Its per capita gross national product was ten times greater than the Philippines', and even slightly higher than Argentina's, although it was less than a third of the United States' or Switzerland's.[6] But poverty is relative: over 90 percent of Italians lacked one or more modern amenities (electricity, drinking water, or a toilet) in their own homes. In the aftermath of war on Italian soil, significant numbers lacked any home at all. High rates of unemployment and of infant mortality were stark reminders of continued economic hardship in a country where over 40 percent of the working population (56 percent in the South) still earned a living — barely — from agriculture. Industry had expanded and modernized under fascism — but only modestly, and only in the North.[7]

From 1945 to 1947, rural insurrection, land occupations and early popular electoral support for the parties of the left raised fears of a drift toward communism.[8] Migrants who had lived long years in the U.S. began a letter-writing campaign to urge their relatives to reject the left. Rather quickly, postwar Italy then fell firmly under American influence.[9] Bolstered by American Marshall Plan funds, and governed by anti-communist Christian Democrats, northern Italian industry recovered from the war's devastation.[10] Southern agriculture did not.[11]

Seeking the loyalties (and votes) of traditionally rebellious but also religious southern voters, the D.C. in 1950 began an important, if uneven, program of agrarian reform to put land into the hands of the poor. Agrarian reform succeeded in small ways and in some areas.[12] But land distribution could not alter the inherent limits of small-scale agriculture on marginal lands. The solutions of the past — especially male migration to jobs elsewhere — still looked attractive to hard-pressed families.

Emigration in the immediate postwar decade was significant — between 200,000 and 300,000 a year, the level of the 1920s.[13] Countries like the U.S., Canada, and Australia emerged economically strong and expansive from the war. Argentina, Brazil, Australia, and Canada again welcomed migrants, and even assisted their migrations. By contrast, the U.S. held firm to its discriminatory quotas until 1965; under its McCarran Walter Act, about 5500 new visas were open to Italians yearly. European nations also continued to link residence and work permits, and to impose a confusing mix of bureaucratic procedures and police surveillance on migrant workers. The liberal assumption that people had a natural right to move about as they pleased had completely disappeared in the face of twentieth-century nationalism.[14]

Unwilling to leave job seekers to fend for themselves, Italy's new government built on older precedents to direct and protect citizens seeking to migrate. It reinstated the independent General Commission on Emigration (which Mussolini had disbanded) but placed the office within the Foreign Ministry, thus treating it (as Mussolini also did) as a dimension of the nation's foreign relations. The decision had a number of consequences. Italy's most important contribution to international diplomacy in the postwar period was its focus on the problems of migration, especially (although not exclusively) within Europe.[15]

Italy's revived labor movement also contributed to state initiatives on migration. Catholics in the D.C. and the C.I.S.L. (Catholic labor federation) continued to see migration as a necessary safety valve, releasing explosive and destructive tensions between rich and poor. Nationalists continued instead to hope emigration could be "a lighthouse of civilization and Great Power of work," as civiltà italiana had been.[16] The P.C.I. and C.G.I.L. (the left labor federation) demanded mainly that the state protect Italy's emigrants from capitalist exploitation abroad. The C.G.I.L. expanded its own initiatives and worked with trade unions in South America and Europe to identify and to propose solutions to the problems of migrant labor.[17]

With labor integrated into state policy-making, the new Italian republic quickly negotiated bilateral agreements that coupled the export and protection of surplus labor to trade, import, and shipping agreements beneficial to Italy as a whole. Typical was a 1946 agreement in which Belgium agreed to sell Italy at a low price 200 kilograms of coal per day for every Italian who went to work in Belgian mines. Between 1946 and 1948, Italy signed similar agreements with France, Sweden, the Netherlands, and Great Britain. Agreements with Argentina, Brazil, Uruguay, and Australia followed in 1948, 1950, and 1951. In some cases, the receiving country or Italy paid transportation costs for migrants; in most cases, the Italian government screened emigrants.[18] The treaties did not challenge any state's power to restrict migration, but highlighted how states "assisted" the migrations they controlled. Three-quarters of Italy's Europe-bound migrants in 1946–54 received state assistance in finding or travelling to jobs.

Multilateral initiatives were less successful. Enthusiastically participating in early steps toward European integration, Italy's ministers pressed the Committee on European Economic Cooperation and the International Labor Organization to create an international migration agency that would prepare standardized regulations for migration worldwide. Because the I.L.O. included representatives from communist countries, the U.S. regularly squashed this idea, and thus protected its highly restrictive immigration policy. Italy also sought free circulation of labor in the European Coal and Steel Community but receiving countries again saw these demands as threats to their national sovereignty.[19]

The padrone state's main goal at home was to diminish social and political discontent among underemployed rural Italians. Abroad, its main goal was to pressure receiving countries to provide welfare services equally to native and migrant workers. This relieved Italy of some of the high costs of birthing, rearing, training, and maintaining in sickness and old age migrants whose labors built other national economies.[20] Italy's bilateral treaties thus helped construct what theorists would later label "social citizenship" — a view that human rights to security and safety exist independent of membership in any particular nation.[21] Because it pursued national interests, too, however, Italy showed little interest in ensuring that migrants had opportunities to settle permanently abroad or to change their citizenship. Both parties to the bilateral treaties of the 1950s anticipated that Italy's workers would return home.

The padrone state concerned itself with male sojourners — as had the earlier padroni. As much as during the mass migrations, agriculturalists most often needed padrone assistance. Comprising 32 percent of Italy's migrants in 1946–55, peasants were better represented than in the interwar years. Male sojourning also became relatively more important than migrations of families during the postwar years. Women's representation dropped slightly to 29 percent — see Table I.4 — and children's representation was lower in the 1950s (6 to 12 percent) than before 1914.[22] The Italian state recruited most postwar male sojourners in the South, where it also concentrated its agrarian reform initiatives. For the first time, more than half of Italy's migrants originated in the South (see Table I.1). In the past, northerners had usually found work across the Alps and southerners across the seas. Southerners now dominated both streams.

Finally, Italy's padrone state was more effective in opening jobs in Europe, South America, and Canada than in the U.S. (see Table I.2). In the first ten years after the war, transalpine migrations (52 percent) surpassed those to the Americas, while Australia became increasingly accessible, and attracted over 5 percent of new migrants. After 1950, migrants to Canada outnumbered those to the U.S., while migrations to South America (excepting only short periods of extremely intense industrial booms in Venezuela and Chile) declined. In Europe, France and Belgium received the largest numbers of workers before 1955.

Italy's padrone state transformed the first phase in the creation of postwar diasporas. It provided many of the same services as padroni, but sought to have bureaucratic ties between individual workers and state agencies replace the face-to-face ties of patronage in home-village communities. Men no longer migrated among fellow villagers to destinations where they found large numbers of paesani who had arrived earlier. The state and its representatives, not the village, became the central node through which information about jobs and wages now passed. State action undermined localism in another way, too, for it gave southerners the familiarity with transalpine Europe which northerners had earlier acquired. Still, sojourners faced a world of harsh restrictions if they considered relocating families abroad. The state did not encourage family relocation, although by the 1950s it was approached by advocacy groups like the Associazione Nazionale Famiglie degli Emigrati (National Association of Families of Emigrants) to examine their problems.[23]

This does not mean that village-based diasporas ceased to exist in the postwar period, but that they emerged mainly through transoceanic migrations, especially newer ones to Canada and Australia.[24] While female representation among postwar migrants dropped overall, Italian women represented over 40 percent of migrants to the

U.S. and Canada — one indicator of families relocating their home bases. (The highly restrictive U.S. accidentally facilitated diaspora-building by granting visa preferences to the close relatives of Italians who had earlier become U.S. citizens.) Rates of return also suggest that new "other Italies" formed across the seas. While rates of return from Europe increased again to two-thirds, those from North and South America declined significantly to 14 and 33 percent respectively.[25]

Typical of the postwar other Italies was Toronto, where over 270,000 Italians lived by 1971. Canada had attracted Italian sojourners in the early twentieth century (see p. 78 above), but Toronto's population of Italians had reached only 14,000 by 1914. Immediately after the war, Canada recruited Italian men as farm workers and laborers, and women as domestic servants. Typical of the laborers was Nino Donato, a peasant who had taken a job in North Africa in 1943, leaving his small plot of Calabrian land to his wife's care. After three years as a prisoner of war, Donato returned to find his family in debt, so he migrated to Canada. By 1950 he had sent for his family and settled in Toronto.[26] Over 40 percent of the 240,000 Italians who migrated to Canada did the same. These "hard-working" people much resembled the mass migrants of the past. Among male workers in 1961, a third of Toronto's construction workers and even higher proportions of its barbers and shoe-repairmen were Italian — but only 7 percent of its factory workers. Half of Italian women workers in Toronto found employment in factories, where they constituted 13 percent of the total female work force.[27]

As families like the Donatos settled in Toronto, the city's earlier Italian migrants were surprised by the continued strength of their village-based diasporas. As in New York at the turn of the century, Catholic national parishes brought together migrants of many regional backgrounds, and their priests could claim with at least some authority to speak for a united Italian community. The heads of Toronto's Italian institutions (especially its mutual aid societies) also expected a shared diaspora nationalism to bind newcomers to their leadership. It did not. Only 6 percent of the new migrants joined existing mutual aid societies. The newcomers formed their own new village-based societies. Since Canada's welfare state provided many of the services such groups had once offered, their functions were essentially social. As workers, Italian men in the construction trades rather quickly became supporters of "bread-and-butter" labor demands, which they supported as militant strikers on picket lines in 1961. But they also at first rejected formal union representation within the Canadian labor movement.[28]

Although significant numbers of new migrants flocked to Buenos Aires, Sydney, and New York in the 1950s, other Italies like Toronto's seemed archaic remainders of an older way of life. In the decade that followed, a new round of sojourner migrations directed to Europe and an economic miracle at home sealed their fate. Fewer and fewer families moved their home bases abroad, eliminating the second phase of migrations linking home villages to their diasporas. New villages of return developed in Italy as sojourners returned from Europe, but the trans-oceanic and village-based diasporas of Italy were near their end.

"Il miracolo italiano": new directions and a new diaspora

In 1957 the Italian state successfully argued that the multilateral Treaty of Rome (the foundation for the European Economic Community) include among its general long-term goals the free circulation of labor within member countries and the equal treatment of workers regardless of citizenship.[29] In practice, little changed in either European migration or citizenship until 1989; restrictions prevailed, as did state control of sojourning. Only two years earlier, in 1955, Italy's government had signed with Germany and Austria precedent-setting agreements that created the "guest worker" migrations that swept Europe in the 1960s.[30] Italy's laborers continued to reorient their migrations away from the Americas and toward Europe.

For the ten years between 1955 and 1965, levels of Italian mobility soared. The number of Italians migrating abroad often surpassed 300,000 yearly. At the same time, observers also excitedly reported an economic boom in Italy's northern provinces that attracted as many of the country's migrants as the expanding economies of northern Europe. Despite encouraging economic signs at home, the Italian government in 1968 entered into yet another round of bilateral agreements (including ones with Germany and Switzerland), ensuring that its citizens could continue to sojourn temporarily in Europe. Immediately thereafter, however, the number of migrants leaving the country plummeted. Northern Italy's "miracle" had been part of a larger European economic boom that both changed the character of Italy's international migrations and ended the country's long history as one of the world's most important exporters of labor.

Italians began to observe signs of an "economic miracle" already in the mid-1950s. It began, unheralded, with the expansion of the automobile, petrochemical, and rubber industries in the early 1950s, with Marshall Fund aid. The end of economic protectionism within Europe allowed Italy's northern steel industry to retool and revive. By the late 1950s, Italian production of consumer goods — washing machines, televisions, typewriters, and refrigerators — had found a strong export market, and the "miracle" began to extend beyond the traditional industrial triangle into the northeast and center. By 1961, 38 percent of Italy's working population were industrial workers, while agricultural employment had declined in ten years to only 30 percent. More Italians, furthermore, could afford to purchase the consumer goods their industries produced. A country that had only 425,000 private automobiles in 1951 had 5.5 million by 1965.[31] Italy's politicians boasted that the economic miracle would end the sharp regional differences that had long fragmented Italy's national economy into unequal segments. "We feel," said one, "that we are resolving the southern problem more than in any other moment of the history of the unified state."[32]

Economic tumult in the late 1960s threatened to end Italy's economic miracle. Climbing unemployment rates in the 1970s, and permanently high unemployment among women and southerners in the 1980s, also suggested the fragility of Italy's national economy. But despite these apparent weaknesses, new medium-sized (and family-based) firms in the northeast and center continued to find good foreign markets into the 1980s. They exported clothing, furniture, household goods, and shoes, specializing in flexible production for markets where tastes changed rapidly. Successful exporters created a "look" that foreign consumers saw as distinctively Italian — a mass, popular, equivalent of the material civiltà italiana of the past. In this respect, Italy's miracle proved more durable than those of northern and central Europe, which had also begun in the mid-1950s.[33]

Responding to the simultaneous development of national and European economies after 1955, Italians rather quickly changed their strategies for pursuing economic well-being. Much like industrializing Germany in the nineteenth century, Italy for ten years was a country characterized by significant domestic and international migrations simultaneously. By the 1970s, it had become both a small-scale sender and a receiver of migrants.[34]

As many as forty-five million Italians (of a total population of fifty million) changed their residence between 1955 and 1981; of these,

Table 7.1 Destinations of mobile Italians, 1955–81 (%)

Destination	*Home region*		
	Northwest	*Northeast and center*	*South and islands*
Northwest	77	10	18
Northeast and center	10	74	11
South and islands	9	5	51
Abroad	4	10	19
(Number in millions	12.1	16.7	16.0)

Sources: Gianfausto Rosoli (ed.), *Un secolo di emigrazione italiana, 1876–1976* (Rome: Centro Studi Emigrazione, 1976), Table 4, and Paul Ginsborg, *A History of Contemporary Italy, Society and Politics 1943–1988* (London: Penguin, 1990), Table 21.

about fifteen million moved long geographic and cultural distances (see Table 7.1). Between eight and nine million of the long-distance migrants moved from Italy's South and center to its more prosperous and industrialized northern regions. Another three million southerners went abroad. Combined, long-distance migrations within Italy and international emigration created levels of mobility comparable to those of the mass migrations.

Domestic migrations from South to North were by no means a new phenomenon. Already in 1901, 203,169 southerners lived in the northern regions of Italy; in 1911 the number had increased slightly to 270,029. World war and restrictions on migration pushed the numbers upward in the 1920s. Migrants obviously defied Mussolini's opposition to urbanization and his plea that peasants remain on their lands to win Italy's "battle for grain" production. In 1931, the numbers of southerners living in the North had more than doubled, to 598,181. A 1939 law then required mayors and urban police to turn away new arrivals and to send unemployed migrants home to their birth villages.[35] While a critic condemned the law as an effort to put Italian citizens under a "sort of house arrest,"[36] migrations continued. In 1951, almost a million southerners lived in northern cities.

In the immediate postwar era, foreign destinations again became more attractive, especially to southerners, and as Table 7.1 also shows, southerners' attraction to foreign jobs remained strong even during the peak years of the economic miracle. Southerners and Sicilians constituted about a third of the internal migrants after

1955, or little more than their representation in the general population. Southerners predominated only among long-distance migrants. Other Italians were equally mobile but stayed closer to home when they moved. While three-quarters of movers from the North and center stayed within their home region, only half of mobile southerners stayed within the South. A long-standing pattern of Italy's migrations thus held: northern and central Italians could find work closer to home; southerners had to look farther away.

Guest workers to Europe were fully three-quarters of Italy's migrants abroad after 1955.[37] The largest numbers went to Germany and Switzerland, which together absorbed 70 percent of Italy's Europe-bound migrants between 1956 and 1965. In the first years after the war, an influx of ten million displaced persons of German descent from eastern and central Europe had met the labor demands of its devastated economy. From 1955 until 1965 Italians became the first, and largest, group among its "Gastarbeiter." Thereafter, Germany signed agreements with Spain, Portugal, Yugoslavia, and Turkey to replace Italians with new recruits.[38]

With its long history of hiring temporary laborers across their common border, Switzerland refused to sign a bilateral treaty with Italy until 1965. Millions of Italians went as sojourners to Switzerland, as they long had, lacking even minimal guarantees of social citizenship. Swiss employers had directly recruited most of them in Italy for temporary, seasonal, and annual contracts. Italians represented 66 percent of Switzerland's foreign workers, and foreigners 10 percent of Switzerland's population even before the bilateral treaty of 1965. Given these numbers, observers in Italy worried about the hostility the next half-million assisted sojourners might face.[39]

The padrone state seems to have delivered Italian workers with at least some urban experience to European employers while leaving less experienced southern peasants to fend for themselves as domestic migrants in Italy's national labor market. The representation of peasants among international migrants declined to only 14 percent after 1955 while peasants made up over a third of internal migrants. Gender ratios were also more balanced among internal migrants, and children better represented — suggesting more frequent family migrations — although even in this group men aged 14–45 generally predominated. Rates of return among internal migrants are harder to calculate, but probably ranged from 10 to 30 percent — lower than among international migrants.[40] When southerners relocated their home bases in the years after 1955, they increasingly moved them to Italy's North, not to northern Europe.[41]

Overall, during the boom years of both Italy's and Europe's economic miracles, Italian industries rather successfully competed with other European nations for the underemployed of Italy's South. As they attracted the southern migrants who in the past went to the Americas, transatlantic migrations dropped off. After 1968 southern migrations to Europe also waned. Scarcely 100,000 emigrants left Italy in the 1970s — the same number that had departed from a much smaller country a century before. Rates of return also soared as the economies of northern Europe soured in the early 1970s. By 1975, more Italians were returning from Europe than seeking jobs there.[42]

As fewer laborers left Italy, the elite migrants who had remained a small component of Italian migrations for centuries again became more visible. Italy's postwar universities continued to produce millions of graduates who faced uncertain futures at home. Italy's industrial miracle did not automatically produce jobs for them. A new journal, aimed at educated migrants (*Italiani nel mondo*), had already appeared in 1945. It promised to report on "Italian work, science, and culture" — e.g. civiltà italiana — but not "in the absurdly imperialist sense it had in the past."[43] It reported on commerce, trade, and migration policy until 1980. That Italy produced a "brain drain" migration after the war's end seems unremarkable. Throughout the postwar world, international migrations generally included a minority (one-quarter to one-third) of well-educated and prosperous professionals and entrepreneurs.[44]

Unlike most brain-drain migrants, however, Italy's mobile elite was not made up of refugees, exiles, or ex-colonials seeking work in the metropoles of their former rulers. Elite migrants chose the same destinations as under fascism: only a third went to Europe, less than a third to the U.S., a quarter to Latin America, and roughly 7 percent each to Africa and Asia.[45] Visual artists like the sculptor Giacomo Scalet in Egypt formed the largest group (26 percent). Italian businessmen and merchants like Michele D'Ambrosio, head of the Industrial Plumbing Company in Toronto, formed the second largest category (20 percent). Scientific and technical workers like Carlo M. Lerici (a specialist in the production of stainless steel in Stockholm) held steady (at 15 percent). The fourth largest group (13 percent) included artists in film, theater, and music.[46]

Unlike the earlier labor migrations with their village home bases, or the guest workers with their padrone state, Italy's elite migrants formed a national diaspora on their own initiative. About five million Italian citizens lived outside of Italy in the 1980s. Already, in 1945 a small group had formed the Association for the Protection of Italians Abroad, and had demanded voting rights and representation for Italian citizens in Europe and around the Mediterranean.[47] Under pressure from labor, the Italian government subsidized "immigrant trains" from several large European cities to allow sojourners to return to vote in Italy's elections in the 1950s. In 1965, the Italian government created and agreed to consult regularly with a committee representing Italians abroad. Most members were Italians working in the many non-governmental and state agencies (patronati) concerned with migrations. Italy's ambassadors represented migrants on the committee. (Only two members of the first committee actually lived abroad; they were editors of Italian-language newspapers.) Discontented elite migrants in the U.S. and elsewhere began to form their own lobbying groups to demand inclusion. In response, the consultative committee ordered the first census of Italians abroad since 1927; registration with the consuls abroad came to be seen as an indication that migrants wanted representation at home. Supporters of the movement noted how it gave Italy "new and more adequate levels of democracy."[48]

Even when they settled in cities like New York, Buenos Aires, or Toronto, Italy's elite migrants had few contacts with the village-based diasporas. In New York, musician Vito Calabrese told interviewers straightforwardly, "My friends are not Italian Americans. . . . It's like two different worlds." Humbler recent migrants agreed with Calabrese. In Brooklyn, Maria Manzone reported the resentment her Italian-American friends expressed at her comfortable life as a recently arrived immigrant. "They keep saying we have it too good. . . . They say they had to wait so long. But it's not our fault. We just came at a different time." Manzone's daughter Angela, a college student, concurred: "People should go to Italy to see how everything has changed. . . . I have an uncle. He's going back after twenty-five years to see what it is like. He still believes there are no washing machines."[49] Italy's economic miracle, together with a welfare state, a consumer society, and the newly visible elite migrants who sold Italian film, technology, and fashion, had redefined Italians as a nation with washing machines and a good head for design. As more Italians stayed home, other nations without washing machines turned their eyes toward Italy as a possible site for their own dreams of economic security.

Guest workers, "immigrants," immigrants and returners

Italy's postwar migrations changed its diasporas, the relationship of northern and southern Italians, and its place in Europe and the world. The vast majority of Italy's migrants had again become male sojourners who returned home in large numbers. Articulate professionals had formed the first national diaspora on their own initiative, and demanded a new relationship to Italy's state. Southerners in Milan became "immigrants" while maids from the Philippines arrived to run washing machines in middle-class Italian homes in Rome. What, if anything, were the consequences for Italy as it confronted not one but four kinds of migration within such a short period of time?

In many respects, Italy's guest workers in Europe reproduced the sojourning lives of the nineteenth-century men without women. Although their wages and working conditions were far better than in the past, their "campaigns" still required most of them to leave families and friends repeatedly for lonely lives at repetitive jobs abroad. In 1965 over two million Italians, most of them from the South and Sicily, labored in Europe, the largest numbers in the four countries of France, Belgium, Germany, and Switzerland. Most were men because industrial expansion had occurred in traditionally male industries like steel, chemicals, and construction. Women made up only 20 percent of Italian workers in France in 1968, and about a third in Switzerland and Germany. The lives of the men without women, as guest workers, remained harsh because the padrone state protected individual workers without challenging barriers to family unification abroad.

Italian guest workers in the 1960s also reproduced some of the old "niches" of the global economy of the nineteenth century. In France, over 40 percent of male migrants worked in construction and public works. The figure was slightly higher in Switzerland. Only in Germany did more Italian sojourners find jobs outside of construction. One-third worked in Germany's metal industries, one-quarter in other manufacturing jobs, and only a quarter in the building trades and construction. At the same time, in France, over 40 percent of Italians were skilled manual laborers, while in Switzerland, semi-skilled factory operatives were the largest group. In Germany, by contrast, almost half of Italians were unskilled manual laborers.[50]

Badly though they wanted guest workers' labor, the receiving countries did not guarantee sojourners a warm welcome. Sojourning gave men little sense of belonging once they left work. In Germany, most working-class natives believed that the men without women chased German women and girls; slightly smaller proportions believed foreign workers started fights or were inclined to violence. Stark evidence showed that guest workers of all backgrounds had higher rates of illness and criminality. Not surprisingly, Italians attracted particular attention as potential criminals. "It's too easy for the foreigners here — paradise for gangsters and criminals," Germany's *Bildzeitung* headlines screamed. Although bilateral treaties guaranteed the sojourners a particular kind of social citizenship, guest workers felt themselves to be — and actually were — socially isolated within the receiving society.[51]

Their integration generally began in national labor movements, where skepticism rather quickly gave way to tolerance and organizing campaigns. As guest workers streamed into the country after 1955, the German Labor Federation (D.G.B.) undertook a program of extensive outreach to foreigners. It provided language and organizational training, as well as again publishing newspapers in foreign languages for workers. The left of France's tripartite labor movement was initially sympathetic to accepting Italian workers in the postwar period, given their positive experiences in a multi-ethnic anti-fascist movement in the interwar years. After a period of resistance to formal recruitment through bilateral treaties, the C.G.T. also quickly began outreach projects. It called for foreign workers to enjoy the same economic and civil rights as natives, including the right to remain in France and become citizens. In Switzerland, by contrast, the largest of the country's many trades unions (the S.G.B., or Swiss Trade Union Federation) opposed recruitment and raised the threat — familiar from the years before World War I — of "Überfremdung."[52]

Ultimately, however, state policy — not the unions — determined guest workers' futures. Switzerland had consistently sought to rotate workers in and out of the country for short periods, discouraging settlement. A large population of daily commuters arrived from Italy, and work permits for residence beyond twelve months were almost unobtainable. (In 1965, only 94,749 of Switzerland's 629,565 foreign-born workers had rights to permanent residence.[53]) As their economic miracles ended between 1971 and 1973, both Germany and Switzerland halted new recruitment, expecting that this would rid their countries of foreign workers. The number of Italians living in Europe declined somewhat between the early 1970s and 1981 but not as much as the receiving states had expected.

The Italian population in France declined from 573,000 to 452,000, and in Switzerland from 587,300 to 417,300. Germany's resident population of Italians, however, only dropped from 629,600 to 624,500. In its 1965 treaty with Italy, Germany had first allowed families to join Italian sojourners. Switzerland, by contrast, still demanded that the Italian government indemnify any welfare costs paid by Switzerland if a settled Italian worker became unemployed.[54]

In the 1970s, European receiving countries discovered — often to their dismay — that Gastarbeiter were in facts "guests who came to stay." A surprising number of sojourners, especially in Germany, had overcome the forest of restrictions to bring families from Italy, or to start new ones abroad. Many "sojourners" had wives, permanent jobs, and children born and going to school in their new homes. Nowhere did national laws provide ready mechanisms for deporting foreign-born workers if they had jobs or legal residence permits. On the contrary, most welfare systems guaranteed foreigners unemployment payments if they lost their jobs. A sizeable population of guest workers in Germany and Switzerland thus became what scholars would soon term permanent "denizens" of the countries where they worked — they were workers who preserved their foreign citizenship while living permanently abroad.[55] The other nations of Europe had helped the Italian state to create firm Italian national identities among such migrant workers.

While Italy's migrants became "denizens" in much of Europe, southerners in Italy's northern cities became "immigrants" in their own land.[56] The new migrants from the South were called immigrants by their northern neighbors with surprising regularity. "Immigrant" lives in the industrial triangle attracted particular attention in the 1960s but "immigrants" had begun arriving earlier, as the figures cited above showed. The automobile manufacturer Fiat, in Turin, for example, had first recruited southern labor in 1925. The impact of such early efforts cast a long shadow. The large-scale industry of the northwest drew many more southern migrants than the newer, smaller-scale industrial areas developing in the northeast and center in the 1960s and 1970s. In work places like Fiat, by the late 1960s, a third of all workers had been born in the South; in some jobs, southerners formed an absolute majority.[57]

The use of the term "immigrants" to describe natives of Italy transplanted from South to North reminds us of the regionalism that persisted beneath Italy's twentieth-century veneer of nationalism.[58] But were southerners as much foreigners in Turin as they were in Berlin or Toronto? It is unlikely. Few southerners needed the assistance of a padrone to find their way to Genoa or Milan but many did seek assistance in finding work there. Called "cooperatives," northern labor agencies specialized in recruiting southerners for short-term, contract labor with employers seeking a temporary work force. Not surprisingly, "immigrant" jobs rarely provided the full range of protections or social security Italian law guaranteed. "Immigrants" worked for lower wages than natives in part because they were concentrated disproportionately in the same kinds of tasks they did abroad — underground work, construction, and — in industry — the janitorial, unskilled, and dirtiest jobs. Like guest workers in Germany, over 40 percent of "immigrants" in the North were unskilled laborers.[59]

Neither did shared nationality eliminate tensions between newcomers and the natives who dismissed them as "caffoni" or "terroni" (rural bumpkins). Keepers of bars and restaurants in districts "invaded" by immigrants insisted that natives and newcomers refused to mingle. Residence patterns also showed signs of segregation. Few "immigrants" found housing in the center of the cities of the North, but concentrated instead in new suburbs, where they sometimes outnumbered natives. While Italy's political parties battled over responsibility for providing new housing for them, the "immigrants" built their own shacks and huts in new, "corea" suburbs. These lacked any public systems, whether transport, water, or sewage. Initially, at least, they differed little from squatter settlements in Latin America's burgeoning cities.[60]

Still, contemporaries tended to overstate the regional and cultural differences between "immigrants" and northerners. By 1970, southern "immigrants" were far more familiar with industrial or urban life than northerners assumed. Surveys of workers in Turin's industries, for example, revealed that between 10 and 30 percent of the southerners had already worked abroad in industry. They had come to Italy's northern cities rather than returning a second or third time to work places in Switzerland, France, and Germany. They had chosen the North over work in Europe because their families could easily join them.

Like migrants abroad, "immigrants" proved a restive but, ultimately, militant work force. Initially excluded from the local labor movement by their temporary and informal jobs, by the late 1960s southerners became a source of renewed militancy in Italy's unions. Some scholars have even attributed the fervor of Italy's "hot autumn" of 1968 to the process of "immigrant" adjustment to industrial life. By joining Italy's labor movement, the "immigrants" confirmed just how Italian they were, and how much like their northern neighbors.[61]

Even in the "immigrant" generation, social interaction with natives outside the work place was also substantial. Perhaps "immigrants" did not mingle with natives in bars, as bar keepers insisted. But if that was the case, then they fell in love in other places. In one Varese (Lombardy) parish, 288 marriages celebrated between 1958 and 1966 (20 percent of all marriages in the parish) involved at least one southern partner. Of these only eighty-four were marriages between two southerners. Seventy-seven joined a southern "immigrant" to another recent migrant from the nearby Veneto. Almost half of the marriages united "immigrants" and natives of Lombardy.[62] Regionalism did not prevent intermarriage; biologically, too, Italy's many peoples were becoming one.

By the time "immigrants" had proved themselves surprisingly Italian, new immigrants were also becoming a source of public concern. Compared to France, Germany, the U.S., or Australia, Italy's immigrant population was modest. The country's 1971 census found only 121,116 foreigners in the country, or 0.2 percent of the population. By 1981, that small proportion had risen to 0.55 percent, or over 300,000 foreigners. By the end of the 1980s, some observers spoke of 600,000 foreigners living legally or illegally in the country, and a 1994 estimate was 899,000 (1.6 percent of the population).[63]

By no means all were "immigrants." In 1981, the largest national groups of foreigners in Italy were from the U.S., Germany, and Greece. Collectively, 168,433 were Europeans and 65,060 North and South Americans; another 4568 came from Australia and New Zealand. Many were students, businessmen, and representatives of foreign governments — part of the same mobile elite that included Italian engineers, artists, and film-makers living abroad. The smaller numbers of labor migrants from the Third World — 41,660 from Asia and 31,571 from Africa in 1981, for example — attracted the only real attention and sparked significant public debates about Italian racism and xenophobia.[64] For a country with a long history of sending migrants abroad, Italy experienced considerable distress in welcoming migrants onto its national territory.

Most of Italy's new immigrants lived in the larger cities of Italy, especially in the North and center, but a surprising minority lived in some of the poorest districts of the Mezzogiorno and Sicily.[65] Just as poorer Polish migrants had poured into the eastern districts of Prussia as Germans emigrated to the U.S. before 1900, so too

Italy's South offered unexpected job niches to migrants. North African street traders replaced native hawkers on Palermo streets and became a ubiquitous feature of southern beaches. Some Third World immigrants claimed to feel more at home in the South. Writing in the 1970s, the Tunisian Salah Methnani, for example, noted: "As soon as I left the central station in Naples, again that sensation of being home, in Tunisia. North Africans everywhere, and everywhere chaos."[66]

Most foreign laborers in Italy were men — but with one very important exception. Women from the Philippines had been among Italy's first immigrants in the early 1970s. Although they fled martial law and underemployment at home to work as domestic servants in Rome, an early study emphasized the women's curiosity about the west, and their affinity with a Catholic country with a long artistic cultural tradition.[67] No studies attributed the migration of Egyptian steel workers, Tunisian street and beach traders, or Sri Lankan domestics to the attractions of the old civiltà italiana. Like the U.S., however, the civiltà italiana popularized by Italy's economic miracle — of Benetton and Versace — was as much part of Italy's attraction to migrants as Coca Cola and McDonald's were in the U.S.

Italy was not universally welcoming to all comers. Like other European nations after the collapse of communism in 1989, Italy vigorously asserted its rights to patrol its borders against unwanted refugees. When 40,000 Albanians fleeing violence in their homeland crowded into boats to cross the Adriatic to Italy's eastern shore in 1991, the Italian military met them. The refugees lived contained within a soccer stadium while the government debated their fate.[68] As this example suggests, Italy faced the same complex issues about the residence, work, and citizenship of foreigners that its citizens had confronted abroad. Italy confirmed its partnership with Europe by adopting familiar restrictions on human mobility. Italian laws passed in 1990 and 1992 distinguished illegal from legal workers (with residence permits). Italy, too, has its illegal "denizens," mainly from Africa and eastern Europe, and their numbers are growing because Italy joined other European nations in issuing work and residence permits very sparingly. Still, immigrants who obtain residence permits can then quite easily join the nation of Italy and quickly acquire citizenship — as many Italian guest workers in Switzerland and Germany could not.[69]

Since Italy's most visible immigrants were from North and East Africa, Italians (like the British and French) talked of their former empire "striking back" at them through migration.[70] Racist and xenophobic reactions to this new immigration differed little in Italy from other European countries. Natives attacked immigrant lodging houses; thugs attacked African immigrants in Florence and in Campania in the late 1980s and 1990s. A nation accustomed to thinking of its migrants as subject to racist and capitalist oppression abroad suddenly looked in the mirror to see itself as the oppressor. Not liking what they saw, Catholic, state, and voluntary organizations sought to improve services to immigrants and to educate native Italians about the positive contributions immigrants made to a growing national economy.[71]

Italy's much larger demographic empire of labor migrants and their descendants also struck back in the postwar years. In 1939, Mussolini had created a permanent commission to aid returning citizens; it continued its work in republican Italy as part of the new welfare state. Beneficiaries of this program came from three directions. Most who benefited were guest workers returning from recent sojourns in Europe. They received financial aid to help them find housing and jobs in their home villages. Basilicata experimented with language courses for the children of returners to prepare them for Italian schools.[72] Calabria noted 64,000 returned between 1973 and 1978, and Abruzzi about 30,000; many had lost jobs abroad. Like earlier returners, the new "germanesi" invested in houses and consumer goods. One returner explained: "When you build your house, you make it for always, and you get yourself all set up in the village for good." Another admitted that a village house still "represents security."[73]

A second group of returners came from the Third World. Just as Germans had fled central and eastern Europe in the years after World War II, considerable numbers of Italians who had relocated to Africa under fascism returned to Italy once its economic miracle began. By 1979 the Association of Italians Returned from Libya even published a newsletter. Joining them were roughly 100,000 children of Italian immigrants who had been born and raised in Argentina and Brazil. They returned to claim Italian citizenship during the "dirty wars" and economic crises of Latin America in the 1970s and 1980s. Attracted by Italy's economic miracle, these returners also reported difficult adjustments. Their dialects were old fashioned. They were no more familiar with modern Italian life than other immigrants.[74]

Italy's awareness of its new status as a receiving country also shaped discussions of relations between nation and diaspora.[75] A world conference for migrants in 1977 focused mainly on the problems of those returning but it also raised the issues that dominated subsequent meetings — notably voting rights through absentee ballot and representation by diaspora deputies in the legislature in Rome.[76] By 1989, some Italian regions were allotting funds to provide cash subsidies for those who wished to return home to vote even in regional elections. Somewhat less than three million of Italy's five million citizens abroad also registered with Italian embassies. In Italy, however, their demands remain controversial. Italy's parties of the left have particularly feared that migrants living in the Americas and Africa might vote for conservatives or even fascists.[77]

Italy's actions toward its many migrants reflect more the country's efforts to bring its policies into line with those of its partners in the European Union than any national accommodation with its diasporas. Italy has not developed a clear understanding of how its history of migration has defined its national identity. As the postwar republic collapsed under charges of corruption in 1992, the country faced harsher tasks than defining a national identity. By comparing the legacies of Italy's many diasporas, however, we can see how much a sending nation like Italy has already come to resemble some other immigrant-receiving nations around the world.

8

CIVILTÀ ITALIANA AND THE MAKING OF MULTI-ETHNIC NATIONS

Tuttu lu munnu è paisi. (Catania, Sicily)
Tutto lo munno è paiese. (Naples)
Totu su mundu est paesu. (Sardinia)
Tutto 'l mondo è 'n paese. (Marches)
Da par to'tt us viv. (Rome)
Tutt al mond è paes. (Bologna)
Tutto u mondo u l'e paize. (Genoa)
L'è töt mond e paìs. (Bergamo)
Tuto 'l mond xe paese. (Venice)
Tut mond è pais. (Piedmont)[1]

A proverb widely cited in Italy, "Tutto il mondo è paese," literally means "All the world is [a] village." The proverb links a cosmopolitan familiarity with the world, il mondo, with the intimate localism of the village, il paese. "Tutto il mondo è paese" sometimes means "All people everywhere are the same." It describes the world as manageable, and as a face-to-face community. The proverb strongly implies that all people can get along, for the world is just a global village.[2]

In the century that has passed since Italian unification, nationalists have taken the word paese (as they have also patria) and partially succeeded in making it mean the nation of Italians and the country of Italy. But this is not what paese means in this proverb. In it, the world and the village are one, and no nation, state, or church mediates between them. The proverb acquires additional layers of meaning when we recall that there is no Italian word for "home." "Tutto il mondo è paese" means that people can feel at home no matter where they are, anywhere in the world. It is a folk statement of a diasporic view of life, appropriate to Italy's migratory peoples and to the village-based diasporas of Italy's mass migrations. Residents of Italy invoke this worldview in ten dialects that represent every region of the peninsula. It is a good expression of the civiltà italiana of Italy's many village-based diasporas.

Forging national solidarity took longer in Italy than in many other parts of Europe, and to this day the existence of the Italian nation remains contested. Having reviewed a millennium of migrations in and out of Italy in previous chapters, I conclude that it could scarcely be otherwise. By looking for an Italian diaspora, this book shows that Italy generated many diasporas but that only one — in the past three decades — was really a "nation unbound." With so many of its residents on the move, Italy's many diasporas complicated its nation-building. They both made the Italian nation and made it what it is — plural, fragile, and debated.

Italy's first diaspora was a trade diaspora of prosperous merchants living in Europe and around the Mediterranean. This diaspora created the concept of "natio" and the related notion that the other Italies of the migrants were colonies, extending the power and influence of their birthplaces. After 1500, Italy's trade diaspora gave way to a cultural diaspora of migratory producers of distinctive forms of art, music, scholarship, and theater. In this cultural diaspora, civiltà italiana marked the residents of Italy as artistic, cosmopolitan, pleasure-loving, and urbane. These were outsiders' views, perhaps, but ones that nevertheless survived Italy's decline into dependency and shaped Italian nationalists' understandings of their nation.[3]

In the modern era, two diasporas of nationalist exiles — radical republican and anti-fascist victims of state oppression — chronologically framed the period of the mass migrations with their thousands of village-based diasporas of migratory proletarians. Both Risorgimento and anti-fascist exiles (like Catholic missionaries, too, and many prominenti) wanted to draw labor migrants out of these diasporas, with their local loyalties, into an Italian nation. They created a variety of diaspora nationalisms in their efforts to "make Italians" abroad; the exiles also shaped the national states created in Italy in 1861 and again between 1943 and 1947. In the name of Catholic and labor internationalism, other exiles drew labor migrants out of their diasporas into more cosmopolitan solidarities, or — more often — into national labor movements and national parishes that opened their path into the nations of the countries where they worked. And, while transnational family economies certainly reinforced Italian localism, they also lured migrant "americani" and "germanesi" back to transform Italy's villages.

Italy's many migrants surely would have agreed with Poland's Marshal Pilsudski that — in the twentieth century at least — "it is the state that makes a nation, not a nation the state."[4] In the interwar years, labor migrants with local loyalties confronted not only Mussolini's demands for loyalty to the Italian "race" at home but demands for loyalty from the multi-ethnic nations where they lived abroad. Mussolini's dreams of empire hung heavily over postwar relations between Italy's state and its migrants. But the padrone state and Italy's economic miracle arguably created real national solidarity for the first time in Italy. Ironically, Italy's last diaspora was also its first national diaspora. Its forgers were surprised to discover their own state reluctant to represent them politically.[5]

Italy's many diasporas allow us to explore the complex relation of cultural diversity and nation-building in the modern world. During the mass migrations, race and class provided the firmest grounds for drawing boundaries around nations. In Italy, the new nation state stigmatized, first, rural peasants and, then, southerners as racially inferior criminals; abroad, receiving states distinguished migrants welcomed as new members of the nation from those rejected or tolerated as working-class sojourners. The result was considerable diversity in employment and labor activism among Italy's migrants, as Chapters 3 and 5 showed. Nation states have reinforced that diversity as they transformed migrants into loyal citizens of multi-ethnic nations.

Today, some countries accept that large numbers of non-citizens will remain permanently settled among their natives. These are the "exclusionary" nations with "ethnic" or "folk" national identities rooted firmly in one cultural tradition. (Germans call them "culture nations.") In such countries, migrants became Italians and denizens. Other countries — the "nations of immigration" — instead acknowledge the culturally diverse roots of their citizens, including the Italians among them. But they take pride in having created nations where ethnic, racial, and cultural distinctions have little place in the public world. Their national identities are "republican" or "unitary." Still other countries celebrate the cultural diversity of their citizens. These are "plural" nations, and their citizens can feel both a sense of italianità and loyalty to the state that governs them as citizens.[6]

Today, identities among the descendants of Italy's migrants differ as much among themselves as migrants once differed from natives. Although diaspora nationalism flourishes among the five million Italian citizens still living abroad, the much larger number of persons of Italian descent do not share it. Migrants' descendants have created their own civiltà italiana based on the thought, "tutto il mondo è paese." Their italianità — where it has persisted at all — resides in the humble details of everyday life, not in the glories of any nation or its state.

Nations of immigration; nations of immigrants; Italians in Europe

The demographic consequences of Italy's global migrations have been enormous. In 1871, less than 2 percent of Italy's twenty-six million residents lived abroad. By 1911 it was 14 percent. By 1920, nine million — a quarter of Italy's resident population – lived outside Italy. The largest group was in the U.S. (five million), with another 2.7 million in South America, and more than a million in Europe. Sixty years later the numbers of italiani nel mondo had declined to five million or about 8 percent of Italy's resident population. Descendants of Italy's migrants are far more numerous than this, however. At sixty million, they about equal Italy's current population. Two-thirds live in the Americas, and most of the rest in Europe and Australia. About fifteen million of Argentina's twenty-seven million residents are of recent Italian descent. As many as 10 percent of French and about 5 percent of Canadian, U.S., and Brazilian citizens are persons of Italian descent.[7] In Italy, too, a sizeable (but unknown) number descend from returners.

Who are these people today? In much of Europe, but especially in Italy, Switzerland, and Germany, the largest group consider themselves Italians, although in Italy significant numbers also identify themselves as Europeans, Lombards, Venetians, or Sicilians. Elsewhere, migrants' descendants are Brazilian, Argentine, and French. In the former settler colonies of the British Empire, more claim hyphenated identities as Italian-Americans, Italo-Australians, or Italo-Canadians. While migrants' paths to national identities were unique in each country, their identities now fall into two types. In France and in Latin America, and in very differing ways in Switzerland and Germany, migrants adopted unitary national identities. In Italy and the former British Empire, plural identities are more common.

Relatively few descendants of Italy's migrants live outside the country in Europe because so many migrants preferred to return to their nearby homes. In addition, Germany and Switzerland usually expected them to do so. Early in the century, the states of Germany and Switzerland initiated a wide range of policies to prevent foreign workers from becoming members of their nations. Like Italy, they chose to define their nations though blood or race (*jus sanguinis*). Later they imposed complex systems to register, survey, and control foreigners' work and residence. In the postwar era, neither Italy, Germany, nor Switzerland saw fit to challenge these exclusionary policies; they shared a common interest in transforming migrants into Italians. Chapters 6 and 7 show how these two countries played an important role in "making" foreign or guest workers into Italians.

In Germany and Switzerland, according to Leslie Moch, foreign workers were "wanted but not welcome."[8] Today, the two countries have extremely low rates of naturalization among resident foreigners.[9] Both continue to insist that they are "not nations of immigrants," leading one cynical scholar to describe Switzerland as "a non-immigration immigration country."[10] Most migrants living in Germany and Switzerland are denizens, and there is at least some evidence that the denizens, too, prefer it that way.[11]

Still, living and working as an Italian is not the same experience in Germany as in Switzerland. Before 1914, Germany used work permits and seasonal rotations to exclude the Polish migrants it viewed as racially inferior but was much less concerned about the work, race, or loyalties of the much smaller migration of Italians. Chapter 5 showed that German labor found creative ways to unionize Italy's workers and to guarantee them equal wages. After World War II, the German union federation D.G.B. helped give guest workers access to the same welfare rights as natives. In Switzerland, by contrast, Italians were the largest group of immigrants and the group that first raised the specter of "Überfremdung." Even in the 1960s, guest workers saw signs warning that the green lawns in parks were off limits to "dogs and Italians." (From the Swiss perspective, this was merely a reminder that people should enjoy the green spaces visually; they were not to be trodden upon, as dogs and outsiders — notably Italians — too often assumed.) "Guests who came to stay" had considerably fewer claims to social citizenship in Switzerland than in Germany, and their social isolation was considerable.[12]

In sharp contrast, Argentina, France, and Brazil developed as self-conscious "nations of immigration." In each, migrants from Italy were important builders of nations with clear, and undivided, national identities. All three countries welcomed migrants from Italy as potential citizens. In all three, a significant minority of Italians settled or worked on the land. In all three, they participated in political struggles that created or recreated the nation. These included the battles of colorados and blancos in the Plata region in the 1840s, the Paris Commune of 1870, and the French anti-fascist movement of the 1930s. Italians were also among the largest immigrant groups in all three countries — 34 percent of immigrants in Brazil and 50 percent in Argentina before 1914, and 40 percent in France in the 1920s and 1950s. Chapter 5 showed Italian immigrants and their children becoming leaders and activists in their multi-ethnic but class-conscious and syndicalist labor movements. In all three, finally, migrants confronted centralizing states determined to create citizens with undivided loyalties.

In Argentina and a few other Latin American countries (e.g. Chile and Venezuela), many migrants from Italy experienced a rapid transition "from exploited to exploiters."[13] Argentina's native, Spanish-origin ruling elite wanted European settlement to push native peoples off the pampas farther into the interior; in cities, immigrants became middle-class employers, businessmen, and industrialists. In Argentine eyes, Italian immigrants brought the technical and cultural modernity of Europe; even their literacy rates were higher than native Argentines'. Italy's migrants could join the nation quickly because their European origins guaranteed them skin color privilege. Native Argentines had rebelled against Spain earlier in the century, and remained sufficiently "hispanophobic" to prefer Italy's migrants to Spain's.[14] Still, Argentina's landed rulers were committed to maintaining their political ascendancy, and uninterested in sharing political power with foreigners or with the poor generally. As Hilda Sabato has written, Argentine politics was "the politics of the streets," not the voting booth.[15] There, too, migrants found incorporation as part of a multi-ethnic "crowd."

Nation-building did not preclude sharp cultural conflicts, especially during the mass migrations between 1890 and 1930. In Argentina, natives dismissed the Italian immigrant as a "cocoliche" or "gringo," and mocked him for his mixed speech. Former migration enthusiast Domingo F. Sarmiento even spoke of Buenos Aires as a "city without citizens . . . a Tower of Babel in America."[16] Diaspora nationalism in Italian institutions in Buenos Aires remained vigorous and influential throughout these years. Still, frictions between Italian and Argentine nationalism also passed quickly. Migrants from Italy soon intermarried with other Argentines. In 1918 in Buenos Aires, half of recently migrated men, and a quarter of Italian women, had married natives or migrants from other national backgrounds.[17]

This bodily mingling encouraged the creation of a new and modern Argentine identity in the 1930s and 1940s. Already during World War I, Robert Foerster claimed that the children of the Italians "are, perhaps, even more vehemently patriotic than the youth of native stock."[18] By the 1930s, the upwardly mobile children of urban middle classes had entered the Argentine military in large numbers. From there, they participated in building populist, corporate, and nationalist movements seeking order after the ten years of political chaos that followed on the "tragic week" of 1919. Their goals were to end the power of British capital in the national economy, to replace imports with textiles and steel produced locally, and to break the power of the old, rural elites. The corporatism of these new nationalists made them attend to the demands of a new industrial labor movement (in which the children of poorer Italian migrants, along with new native-born arrivals from the countryside — the descaminados — continued to be visible leaders). Like the elite that preceded them, however, Argentina's nationalist military had little interest in building a democratic polity or a modern welfare state.[19] By the early 1940s, Argentina's labor movement became the power base for Juan Peron (himself the son of an Italian mother), and for the populist and militarist economic nationalism of Peronism.[20]

Even more than Argentina, Brazil's creole elites sought European migrants to "bleach" a nation that blended European, African, and indigenous peoples. Brazil's rulers, too, claimed to prefer migrants from Italy to those from Portugal, because they lacked the "arrogance" of former colonizers. Some even argued, "there is nothing more Brazilian than a first-generation Italian."[21] Brazil's policy worked: in 1870, Brazilians of European descent were a third of the population; by 1940 they were over half, and in cities like São Paulo even more.[22] As in Argentina, there were cultural conflicts: native Brazilians referred to Italian immigrants dismissively as "carcomanos." Overall, nation-building in Brazil followed much the same route as in Argentina. A new nationalist movement in the 1930s excluded foreign capital, encouraged industrialization, and centralized political power in Brazil's most Europeanized cities. The descendants of Italians participated in these transformations.

In Europe, Italian immigrants developed French identities via a different path. France's slow-growing population and revolutionary tradition kept it a nation of immigration even during periods of intense cultural conflicts between French workers and the migrant "ritals" and between France and Italy. (The Aigues Mortes massacre had occurred during a French–Italian trade war.) Italians entered the French nation quickly via the country's multi-ethnic, syndicalist, and anti-fascist labor movement. Rates of intermarriage were as high as in Argentina, and in the 1970s they had reached 80 percent.[23] Rather than blending to create a new national identity, as occurred in Argentina and Brazil, however, France succeeded in imposing Franco-conformity on its many migrants.

A key to its success was an activist state committed to transforming migrants, as it had earlier changed "peasants into Frenchmen."[24] Beginning with its republican revolution — which made citizenship voluntary and open to any man — the French state drew sharp civil distinctions between citizens and foreigners but totally ignored — and even denied — racial, regional, religious, and ethnic divisions among its citizens. Migrants could naturalize after only two to five years of residence, although they waited longer before attaining voting rights. After 1889, the French-born children of Italians in France could also easily claim French citizenship at age eighteen (if they still lived there). To encourage foreigners to become French, the state created privileges for citizens: foreigners could not hold civil service jobs, practice many professions, or run certain small businesses. They could belong to unions but neither hold high office in them nor direct them. Beginning in 1914, France required foreigners to register with the police and to carry special identity cards, and it limited welfare rights to foreigners who continued to reside in France.[25] According to Gérard Noiriel, French "citizens control the state via their representatives . . .; on the other hand, the state directly contributes to the making of personal identities by codifying the main elements that define them."[26] A separate and distinct sense of italianità would not be one of those elements — as it was elsewhere.

Like the "nations of immigration," the "nations of immigrants" — the U.S., Canada, and Australia — wanted immigrants. But unlike their Latin counterparts, these largely Protestant countries dominated by English-speaking and Anglophile elites did not welcome Italians, and looked at all Catholic and Jewish immigrants with considerable concern. Italians arrived too late, and too poor, to farm their frontiers or contribute to nation-building, as "old stock" settlers had. Worse, natives viewed Italians and other southern and eastern Europeans as racially ambiguous "in-between people."[27] Neither white nor black, Italians were "the Chinese of Europe" in all three countries. Marginalized in sojourning jobs, excluded from the countries' labor movements as radicals before World War I, and unable to enter freely during the interwar years because of racially restrictive laws, immigrant Italians were tiny minorities in these multi-ethnic populations. The U.S., Australia, and Canada wanted them as workers (as did Switzerland) but hesitated to welcome them as new members of the nation. All restricted migration from Italy at times. Migrants who settled in the U.S., Australia, and Canada found their entrance into the nation hedged by many obstacles; experiences of discrimination, coercion, and fear accompanied nationalization. Plural or "hyphenated" identities were the most common result.

As a self-proclaimed nation of immigrants, the United States provides the clearest contrast to republican France. Unlike France's revolutionaries, the "founding fathers" of the U.S. acknowledged regional differences (notably slavery) in their Constitution. Until 1964, discrimination against African-American citizens was legal under many local laws. The federal and decentralized government of the U.S. also allowed its citizens to organize churches and regional, racial, and ethnic associations unimpeded, and all of these became involved in electoral politics, both directly and indirectly. As cultural conflicts grew during the mass migrations, stigmatizing Italian migrants as "wops," "dagos," "guineas," "Black-handers," and mafia criminals, the segregation of Italian immigrants in the U.S. seemed quite pronounced.[28] Chapters 4 and 5 revealed ethnic segmentation in the labor movement, the Catholic church, and in urban "Little Italies." America's laissez-faire state did not create this segregation but rather tolerated it as a voluntary creation of local communities, and thus of no concern to the federal government. Chapter 6 showed that migrants entered the American nation as Italians, and as Catholics, with an "ethnic" and religious label rooted in the competing diaspora nationalisms of fascism and anti-fascism.

However hostile to migrants on racial grounds, the U.S. (like Brazil, Argentina, Canada, and Australia, too) granted citizenship automatically to all children born on its soil (*jus soli*). It accepted as citizens the children of even excluded migrants (like the Chinese) or restricted migrants (like the Italians). Voting and participation in electoral politics ultimately became a key nation builder in a country where universal manhood suffrage had preceded the mass migrations. Like France, the U.S. facilitated easy naturalization (within three to five years of residence) for European (although not Asian) immigrants. But political incorporation, too, proceeded through ethnic and religious channels.

Early in the century, padroni led their clients into the urban political "machines" for which American cities became famous. In New York, for example, James March (Antonio Maggio) first sought leadership in the Irish-dominated Tammany Hall machine; failing at that, he became a Republican, as did many new Italian citizens thereafter. In the 1930s, the progressive Fiorello La Guardia — a staunch supporter of Roosevelt's New Deal welfare state — was a sometimes-Republican, as was the leftist radical Vito Marcantonio in the postwar era.[29] Migrants and their American-born children organized Italian-American clubs within both parties. To this day, politicians of Italian descent, like the Democrats Mario Cuomo and Geraldine Ferraro or the Republican Rudolph Giuliani, are assumed beneficiaries (without much evidence) of an Italian "ethnic block" of voters.[30] Ferraro, in particular, has been questioned repeatedly about her family's ties to organized crime. The salience of these political leaders' Italian roots contrasts sharply with their irrelevance in the far more prominent and long-lasting career of Argentina's Juan Peron.

The consolidation of loyalties to the American nation through male political participation outpaced cultural and social incorporation. The U.S. had a weak welfare state and a labor movement uninterested in the social citizenship of foreign workers. In 1918, furthermore, when high proportions of Italians in Buenos Aires were intermarrying with other migrants and with natives, three-quarters of Italian men in New York, and all Italian women, married other Italians.[31] Marital amalgamation increased among migrants' children. As many as half chose partners from other recently arrived Catholic immigrant groups; marriages with native, Protestant, and "old stock" Americans were fewer. The occupational status and education of Italian-Americans lagged behind other European immigrants until the 1960s when the third generation — migrants' grandchildren — entered college.[32]

Today, there are few remnants of Manhattan's many Little Italies, and these are mainly clusters of businesses aimed at the tourist trade, along with the churches that were once the center of national parishes. New York's Lower East Side Little Italy has almost disappeared in an expanding Chinatown; East Harlem is Black and Hispanic. There are Italian-American residential clusters in New York's other boroughs, however, and in the suburbs of greater New York, too.[33] Italian-Americans have not spread through the country but still cluster in the northeast and around New York City. The majority of Italian-Americans are three or four generations removed from the migration experience. A few weekly newspapers publish in Italian, and the Sons of Italy survives as a federation of mutual aid societies. Educationally and occupationally, Americans of Italian descent do not differ much from other urban Americans in the northeast. Richard Alba's studies suggest furthermore that as many as three-quarters are the products of ethnically mixed marriages, although intermarriage among Catholics remains important.

Still, as Alba also argues, Americans with hybrid roots more often chose to remember the "Italian" than other cultural origins. "Italian" is also one of the few ethnic markers persistently listed in personal ads posted in U.S. newspapers. More than half of migrants' grandchildren still call themselves "Italian-Americans" (as does this author). With few of the vestiges of the old village-based diasporas remaining, such plural identities seem puzzling. Many scholars emphasize the ambivalence they symbolize.[34] Migrants discovered, says one, that the U.S. changed "from promised land to bitter land."[35] Another study claims Italian-Americans are "at home and uneasy in America."[36] Dino Cinel offers a provocative explanation; he argues that Italian-Americans may feel they failed because they never attained their goal of return to the village.[37]

In Australia and Canada, too, migrants developed hyphenated identities. These countries shared the "Italophobia" that Robert Harney termed "an English-speaking malady."[38] Still, both Canada and Australia abandoned discriminatory immigration restrictions long before the U.S., and they welcomed many more migrants from Italy in the postwar era, as Chapter 7 showed. The product of large, more recent migrations, Italo-Canadians and Italo-Australians are younger; more are of the first and second generation and have closer personal ties to modern Italy and to their village-based diasporas. More important, however, in creating new national identities, migrants to these countries confronted states that actively promoted multicultural public policies in which Italians had a prominent place.

Canada, of course, had been bicultural and bilingual since the seventeenth century. Its state-sponsored celebrations of multiculturalism began in the 1970s as part of an effort to defuse growing tensions between separatist French-Canadians and the English-speaking Canadians who had shown little patience with cultural diversity in the past. Italo-Canadians became one among several dozen Canadian ethnic groups acknowledged and celebrated as part of a new Canadian plural nation. Plural identities became something all Canadians shared; they united, rather than divided, the nation.[39] The same was true in Australia. There, state recognition of Italians' contributions as a once-disparaged and excluded European group became a first step away from notions of a "White Australia," deeply rooted in the imperial past. In Australian celebrations of multiculturalism, the recruitment of Italian workers after World War II marked a significant departure from the racism of the past. It opened subsequent possibilities for welcoming and incorporating Asians and other new migrants.[40]

The plural identities of Italo-Australians and -Canadians have a firm place in these countries' understandings of their multi-ethnic nations. In the U.S., by contrast, the nation seems unsure of where Italian-Americans belong in the multi-ethnic rainbow. Some scholars deny that Italian-Americans are a culturally distinctive group at all, and they dismiss their hyphenated identities as a symbolic or romantic atavism among a people that is firmly Euro-American, white, and part of the privileged mainstream.[41] In the U.S., multiculturalism focuses more on ending the exclusion from the nation of minorities recognized officially in state-mandated affirmative action categories. These programs, which originated in the Civil Rights legislation of the 1960s, recognized the systematic racial and cultural exclusion experienced by Hispanics, African-Americans, Native-Americans, and Asian-Americans at the hands of white Euro-Americans. Italian-Americans' continued embrace of the hyphen and of a plural identity may indicate just how "in-between" many still feel in this debate, even in a nation where — as one former critic noted — "we are all multi-culturalists now."[42]

Italian-Americans share their preference for plural identities with their distant cousins in Italy. The collapse of Italy's postwar republic in 1992 — a political earthquake, commentators called it — revealed both the persistence of regionalism and the popularity of a cosmopolitan internationalism among present-day residents of Italy.[43] Italy's state had consciously begun a course of devolving responsibilities onto its regional governments in the 1970s, thus increasing regional and local autonomy. But the revelations of widespread corruption among Italy's leaders in Rome, regardless of regional background, also seemed to belie old notions of divisions between the clientelistic and mafia-ridden southerners and civic-minded northerners.[44] In the maelstrom of rewriting Italy's Constitution and reforming its state, many Italians in the prosperous North (where incomes are still 50 percent higher than in the South) nevertheless organized a separatist Northern League. It called for independence for their state of Padania (the lands north of the Po River). At the same time, Italy's population includes higher proportions of convinced "Europeanists" and supporters of European unity than any other country in Europe. Many Italians welcome prospects of the cosmopolitan future the European community promises. Others believe that only Europe can release their nation from the stranglehold of corrupt parties, cynical party leaders, and the musical chairs of coalition governments.[45] For Italy's "Europeans," a critic claims, "Europe has become a priest; getting into EMU [the European Monetary Union] is like going to confession."[46]

Nation-building eradicated diaspora nationalism in some countries, and facilitated it in others. Argentina, France, and Brazil believe they created melting pots — "le creuset français" or "lo crisol de razas" — that ended the divisions of diaspora nationalism among their immigrant populations.[47] They recognize that their peoples have origins elsewhere — as in the joke that claims "Argentines are Italians who speak Spanish who think they are British."[48] Latin Americans also acknowledge ties to Italy as sentimental and familial, as did the immigrant in Mexico who maintained, "America is our mother, Italy is our grandmother."[49] And the French sometimes speak of their "populations d'origine immigrée [populations of immigrant origin]." But origins and family ties are unconnected to national loyalties in these countries; in their national identities, migrants' descendants are simply French, Argentine, and Brazilian. We cannot even trace the assimilation of the second-generation children of immigrants in these countries because their states keep no public records that indicate origin, nativity, or religion.

As one student of French policy has noted, "It is possible to be an Italian in France, but it is not possible to be an Italian-Frenchman in the same way it is possible to be an Italian-American."[50] In the past ten years some French and Argentine intellectuals have grappled with their nations' historical experiences with racism, and considered the possibility that there have also been French and Argentine forms of pluralism. The French, in particular, have been polarized by a xenophobic nationalist movement alarmed at the rising number of dark-skinned Frenchmen coming from France's ex-colonies in the Caribbean and North Africa.[51] Both France and Argentina, however, view multiculturalism as a failure of nation-building and a product of racism in the English-speaking world.[52]

In most other places where Italians settled, pluralist nation-building has been the norm, but it has assumed very different forms.

When Germans discuss multiculturalism, many imagine people remaining members of their own "culture nations," no matter where they live or work, yet being able to live and work together in peace. The possibility of a plural but federal European polity reinforces such views. Some Germans even argue that *jus sanguinis* and social citizenship offer denizens (and especially their children) the greatest degree of cultural freedom by allowing them to remain citizens of their nations of origin if they choose. (This is a choice that American *jus soli* forecloses.)[53] By contrast, when English-speakers invoke a multi-ethnic nation, they imagine peoples with a variety of cultural identities sharing loyalty to one national state and to its particular political system. In Italy, unsurprisingly, the plural nation is one of peoples whose identities are both regional and cosmopolitan, making room for the new immigrants who increasingly live among them to develop their own plural, and probably ethnic, identities. Italy has become a "nation of immigrants" like the U.S. or Australia in this other sense, too.

Whose civiltà?

Comparing peoples who migrate to different places inevitably emphasizes their differences.[54] If we instead ask what, if anything, the descendants of Italy's migrants share worldwide, we also ask, in effect, about the survival and transformation of civiltà italiana over the centuries. Early chapters showed that each of Italy's diasporas created cultural, political, and social forms which their producers, and outsiders, understood to be Italian. Art, music, and scholarship were the products of the earliest Italian diasporas. Garibaldismo, rebellious radicals and conspirators, Italian ices and street music were products of the diasporas of the Risorgimento. Italy's most recent diaspora has created its own version of civiltà italiana. As a glossy magazine for diaspora Italians asked in its inaugural issue,

> Does it make sense to call oneself or to feel Italian on the threshold of the third millennium? Differentiating ourselves from regurgitating nationalists, we firmly believe that it does. . . . There is an Italian style, an Italian culture, an Italian capability of which we can be proud.[55]

In this last diaspora, civiltà italiana may mean Milan's sofas or menswear, the leatherware of Fendi, fine Tuscan food, or a generally romantic and hedonistic zest for living. It is a modern, and corporate, version of the urban pleasures of the Italian "style" of the Renaissance. This civiltà italiana has a commercial presence throughout the world.

But it is not the civiltà italiana of the largest of Italy's diasporas, the ones formed by the workers of the world as they traveled the globe ceaselessly in search of economic security. Most persons of Italian origin in the world today are descendants of these humble migrants. In Europe, and in the former British Empire, where Italian identities persist, civiltà italiana includes, instead, defensive reactions to the negative stereotypes with which the Italian state marked rural Italians in the nineteenth century. It reveals a continued lack of interest in states and nations, a love of the pleasures of everyday life, a commitment to local face-to-face communities of families and neighbors, and a continuing involvement in Roman Catholicism.

Of these, indifference to citizenship is the easiest to document and hardest to prove — at least in those countries that welcomed, or at least tolerated, Italy's migrants, as potential new members of the nation. In Argentina, before World War I, for example, fewer than 2 percent of Italian migrants had acquired citizenship (or "naturalized") in their new homes. Changes in citizenship were equally rare in Brazil. That country created citizens by imposing it automatically on migrants who married natives, bought land, or made important contributions to business and commerce. After World War II, Argentina, too, automatically naturalized foreigners who had lived in the country for five years.[56]

Of course, citizenship in Argentina and Brazil carried few rewards to tempt migrants: in the early years of the century, few citizens could vote, and even after 1912 Argentina linked men's voting rights to completion of military service.[57] Until the 1980s, in Argentina, voting had little relevance in a country dominated by the military. But even in the U.S. and France, where universal manhood suffrage was the norm and an important means to enter the nation, naturalization rates among Italians lagged behind other immigrant groups. In the U.S. in 1911, for example, about a third of other foreigners had acquired citizenship, but only 12 percent of New York's Italians had naturalized. Proportions were even lower in France. By the 1920s, these rates had increased to about a quarter in the U.S. In France in the late 1930s, they soared — but even then only to 30 percent (which is about where they remained in the postwar era).[58] Reflected in such figures are migrants' expectations of return, as well as echoes of rural cynicism toward "governo, ladro."

Although relatively few migrants changed their citizenship and many clung to their native dialects, few of their descendants speak Italian today.[59] Still, the migrants left a linguistic mark where they settled in large numbers. In Buenos Aires, Spanish-speaking visitors claim to hear the lilt of the Italian past in the "porteño" accent of the city's multi-ethnic population. Some New Yorkers, too, claim they can distinguish variations in the "Big Apple" accents of English speakers of Italian, eastern European Jewish, and Irish descent. But these are isolated examples. Worldwide, Italian migrants have most often contributed to other languages the terms of their own disparagement and isolation: ritals, carcomanos, gringos, wops, and dagos. Even more noticeably, they have given Europe and the Americas the concept (if not always the phenomenon) of mafia.

The criminalization of Italy's migrants originated in Italy's post-unification civil wars between state and peasantry, as Chapter 2 revealed. Charges of criminality (along with anarchy and violence) then followed Italy's migrants everywhere, generating both long-lasting stereotypes and a vocabulary of crime that emphasizes its origins in Italy. "Camorra," "cosa nostra," and other local terms for criminal protection rackets gradually gave way to the more universally used "mafia."[60] But even mafia is more closely associated with Italians by English and German speakers than by those who speak French and Spanish. Indeed, the role of Italians in organized or "big-business" crime may have been greater in the U.S. than elsewhere. As one Italian-American, frustrated with the constant association of crime and Italians, noted, "Sicilians have gone all over the world, Argentina, Brazil, Australia, all over and the only place where this mafia developed was here in America. Can you tell me why?"[61] Ironically, the U.S. film industry — beginning with Jimmy Cagney playing an Italian in gangster movies, and continuing in the work of Italian-Americans from Mario Puzo to Martin Scorsese — have spread these American versions of old Italian stereotypes throughout the world.[62]

In Germany, natives sometimes refer to pizza as a "mafia torte" (mafia cake), thus bringing together the two most important symbols of labor migrants' civiltà italiana. Italian migrants — many of whom had never eaten any of these things at home — brought ice cream, pizza, and spaghetti to most of Europe, to Buenos Aires and São Paulo, to Toronto and Sydney, to New York and San Francisco.

The pizza of Buenos Aires is different from that of New York (or Chicago). But milanese alla napolitana (made with Argentine beef) and veal parmigiano (a standard in South Italian restaurants in North America) share visible roots in the wienerschnitzel of Austrian-dominated Milan.

Throughout the world, natives have associated with Italy's migrants the love of good, bounteous food, washed down with "dago red" (as it was called in New York at the turn of the century), Chianti or Marsala. Even the French — notorious preservers of their own culinary traditions — have accepted the noodle dishes and pizzas of Italy. (Indeed it was a French immigrant who first introduced spaghetti alla milanese to American consumers in a Franco-American can in the 1890s.) In Germany today a "spaghetti-fresser" (someone who eats large quantities of spaghetti) is as often a native as a foreigner, and Turkish guest workers in Berlin now operate pizzerias where döner kebab and kefir jostle "pizza Hawaii" — clearly an American specialty — on the menu.[63]

In the German- and English-speaking worlds, love of food, intense and warm family lives, patriarchal restraints on women, and religious traditionalism are all traits attached to those of recent Italian origin.[64] While Latin Americans tended to view their Italian immigrants as "modernizers," these images instead cast civiltà italiana as either backward and conservative or more positively rooted in premodern ways. In the U.S. and Canada, Italian-Americans themselves celebrate the uniqueness of their family loyalties.[65] Many view their family loyalties and their religious convictions as healthy alternatives to modern moral decline, the anonymity of modern cities, the loneliness of the marketplace, and the individualism of modern American Protestantism.[66] In the U.S., some Italian-Americans also see family ties and Catholicism as bulwarks against the rampant individualism of Anglo-American culture or the best mechanism for differentiating themselves positively from African-Americans or more recent immigrant arrivals.

In the English-speaking world and in Italy, localism among migrants' offspring is also part of the long-term cultural legacy of Italy's proletarian but village-based diasporas. Whether in the form of turf battles with new immigrants and African-Americans or a preference to live near relatives, a peculiarly Italian love of the patria remains an important continuity in civiltà italiana in the U.S.[67] In New York, Italian-Americans remain in Brooklyn and the Bronx while other immigrant groups leave in larger numbers for the suburbs. In Italy, middle-class professionals may travel weekly by overnight train to distant jobs in order to maintain residences in their hometowns. Surely, there is at least a slim cultural thread that connects these choices.

Just as surely, they represent a skepticism about the nation as Italy's nationalists first imagined it. Even Mazzini had insisted that "La patria isn't a place; the territory is nothing more than the basis. The patria is the idea that emerges from it: the sense of love and communion that pulls at all the children of that territory."[68] In the modern civiltà italiana of Italy's proletarian diasporas, home is still a face-to-face community — not the idea of a people rooted to a place, but the place itself. And that place can be anywhere in the world; it is not necessarily Italy, but a well-known village, neighborhood, or city anywhere in the world.

The civiltà italiana of the proletarian diasporas seems at first glance a defensive culture, and largely a product of migrants' experiences in the English-speaking world. It is as ambivalent, as burdened with the negative stereotypes of criminality, and as humble as most of the unskilled southerners who flocked to the U.S. Of the oldest diaspora cultures, Catholicism has the strongest influence on the descendants of migrants today — especially in the primarily Protestant receiving countries of the English-speaking world — while the arts and music of earlier diasporas are of considerably less importance. The labor militancy and internationalism of the proletarian diasporas appear to have left little mark; their major accomplishment was to transform migrants into Argentines, Brazilians, and Frenchmen.[69] The bombastic claims fascists made about Italian technical and racial superiority and their imperial commitments to a civilizing mission in the Third World have left few traces among migrants' descendants today. The civiltà italiana of Italy's proletarian diasporas was a surprisingly modest one. Ever aware of its humble and often disparaged roots, it nevertheless still offers an important alternative to all modern nationalisms, even those of Italy. It continues to emerge from the local ties — and the everyday pleasures — of food, family, parish, and home place, all things that can be enjoyed and savored anywhere in the world that people call home.

NOTES

INTRODUCTION

1 Walter F. Willcox and Imre Ferenczi, *International Migrations* (New York: National Bureau of Economic Research, 1929), vol. 1, p. 811.

2 The by now well-known concept of "imagined community" is from Benedict Anderson, *Imagined Communities: Reflections on the Origin and Spread of Nationalism* (New York: Verso, 1983).

3 Giulio Bollati, "L'italiano," in Ruggiero Romano and Corrado Vivanti (eds), *Storia d'Italia*, vol. 1: *I caratteri originali* (Turin: Giulio Einaudi, 1972), pp. 951–1022. Here Bollati refers to the early commentary of Vincenzo Gioberti, writing in 1846 on the absence of an Italian nation or people. For a modern discussion of the problem, see Giorgio Calcagno (ed.), *Bianco, rosso e verde: L'identità degli italiani* (Rome: Laterza, 1993).

4 Robin Cohen, *Global Diasporas: An Introduction* (Seattle: University of Washington Press, 1997), p. ix. Cohen's work provides both a typology of diasporas and an excellent introduction to the multi-disciplinary theoretical literature on diasporas. See, e.g.: J.A. Armstrong, "Mobilized and Proletarian Diasporas," *American Political Science Review* 20, 2 (1976): 393–408; James Clifford, "Diasporas," *Current Anthropology* 9, 3 (1994): 302–38; Stuart Hall, "Cultural Identity and Diaspora," in J. Rutherford (ed.), *Identity: Community, Culture, Difference* (London: Lawrence & Wishart, 1990); Khachig Tölölyan, "The Nation-State and its Others: In Lieu of a Preface," *Diaspora* 1 (1991): 3–7; William Safran, "Diasporas in Modern Societies: Myths of Homeland and Return," *Diaspora* 1 (1991): 83–99.

5 The best summary of numbers for the modern, national, period is Gianfausto Rosoli (ed.), *Un secolo di emigrazione italiana, 1876–1976* (Rome: Centro Studi Emigrazione, 1978). I have taken most of the raw figures on the numbers and characteristics of Italy's migrants from this source, which is based on Italian government statistics.

6 I say "roughly" because the collection of data on border-crossing has been an irregular concomitant of nation-state formation. Numbers are firmest for trans-oceanic migrations from Europe; between 1830 and 1930 they equaled roughly sixty million. Within Europe, states did not always register international and long-distance migrations before 1914. However, it is likely that they surpassed the trans-oceanic migrations in volume. On the former, see Walter Nugent, *Crossings: The Great Transatlantic Migrations, 1870–1914* (Bloomington: Indiana University Press, 1992).

On the latter see Leslie Page Moch, *Moving Europeans: Migration in Western Europe since 1650* (Bloomington: Indiana University Press, 1992), ch. 4. In addition, approximately thirty-nine million Asians (thirty million Indians; nine million Chinese) emigrated during the same period, two-thirds of them within Asia. Kingsley Davis, *The Population of India and Pakistan* (Princeton, N.J.: Princeton University Press, 1951), pp. 13–14; Victor Purcell, *The Chinese in Southeast Asia* (London: Oxford University Press, 1951), pp. 43, 85, 175. The total number of international migrants during this 100-year period was probably not much less than 150 million. Lydia Potts, *The World Labour Market, A History of Migration* (London: Zed Books, 1990), ch. 3.

7 The localism and regionalism of Italy's residents are much cited. English speakers will want to start with Denis Mack Smith, "Regionalism," in Edward R. Tannenbaum and Emiliana P. Noether (eds), *Modern Italy* (New York: New York University Press, 1974), and Carl Levy (ed.), *Italian Regionalism: History, Identity and Politics* (Oxford: Berg, 1996). See also Giovanni Levi, "Regioni e cultura delle classi popolari," *Quaderni storici* 15, 2 (1979): 720–73. Much but not all of this literature focuses on the "southern problem." Robert Lumley and Jonathan Morris (eds), *The New History of the Italian South; The Mezzogiorno Revisited* (Exeter: University of Exeter Press, 1997). Americanists, by contrast, more often invoke "campanilismo" — the devotion to the local church tower, parish, or immediate neighborhood — but no critical or scholarly examination of this term exists.

8 Maurizio Viroli, *For Love of Country; An Essay on Patriotism and Nationalism* (Oxford: Clarendon Press, 1997).

9 Both terms became popular in the aftermath of World War II as scholars rejected an earlier scholarly discourse that viewed Italy's diasporas as "colonies"— a term that had become tainted by fascist imperialism (see Chapter 6). See Carlo Morandi, "Per una storia degli italiani fuori d'Italia (A proposito di alcune note di A. Gramsci)," *Rivista storica italiana* 3 (1949): 379–84; Varo Varanini, "Gli italiani nel mondo," in Corrado Barbagallo (ed.), *Cento anni di vita italiana 1848–1948* (Milan: Cavallotti, 1948), vol. 1, pp. 495–536. See also the periodical *Gli italiani nel mondo* (1945–80).

10 Jacques Le Goff, "L'Italia fuori d'Italia: L'Italia nello specchio del Medioevo," and Fernand Braudel, "L'Italia fuori d'Italia; Due secoli e tre Italie," in *Dalla caduta dell'Impero romano al secolo XVIII*, and Robert Paris, "L'Italia fuori d'Italia," in *Dall'unità ad oggi*, vols. 2 and 4 of Romano and Vivanti (eds), *Storia d'Italia*.

11 Ernesto Ragionieri, "Italiani all'estero ed emigrazione di lavoratori italiani: Un tema di storia del movimento operaio," *Belfagor, Rassegna di varia umanità* 17, 6 (1962): 640–69; Bruno Bezza, *Gli italiani fuori d'Italia, Gli emigrati italiani nei movimenti operai dei paesi d'adozione 1880–1940* (Milan: Franco Angeli, 1983). Following Willcox and Ferenczi, in *International Migrations*, Dirk Hoerder prefers the term "proletarian mass migration," in his "An Introduction to Labor Migration in the Atlantic Economies, 1815–1914," in *Labor Migration in the Atlantic Economies; The European and North American Working Classes During the Period of Industrialization* (Westport, Conn.: Greenwood Press, 1985),

pp. 6–7. See also Donna Gabaccia and Fraser Ottanelli, "Diaspora or International Proletariat?," *Diaspora* 6, 1 (1997): 61–84.

12 Probably the first scholar to use the term (which he placed carefully in quotation marks) was Morandi in "Per una storia degli italiani fuori d'Italia." More recently see George E. Pozzetta and Bruno Ramirez (eds), *The Italian Diaspora: Migration across the Globe* (Toronto: Multicultural History Society of Ontario, 1992); Gianfausto Rosoli, "The Global Picture of the Italian Diaspora for the Americas," in Lydio F. Tomasi, Piero Gastaldo, and Thomas Row (eds), *The Columbus People: Perspectives in Italian Immigration to the Americas and Australia* (New York and Turin: Center for Migration Studies and Fondazione Giovanni Agnelli, 1994), pp. 305–22. Although not using the term diaspora, the earliest comprehensive study of Italian migrations also took a global approach: Robert Foerster, *The Italian Emigration of Our Times* (New York: Russell & Russell, 1968) (first published 1919).

13 Cohen, *Global Diasporas*, p. 2. About 11 percent of the migrants described in a biographical dictionary of elite Italians were exiled or banished during the 1300s, usually as a result of factional fights within their city states. The same source suggests that 28 percent of elite migrants between 1790 and 1860 migrated for political reasons. Ugo E. Imperatori, *Dizionario di italiani all'estero (dal secolo XIII sino ad oggi)* (Genoa: L'Emigrante, 1956).

14 My evidence on the earliest migrations from Italy, 1200–1789, as well as elite migrations in the modern period, is based on systematic analysis of a biographical dictionary of Italy's most prominent migrants. See Imperatori, *Dizionario di italiani all'estero (dal secolo XIII sino ad oggi)*, which contains short biographies of 3200. (For a discussion of Imperatori's work, see Chapter 1.)

15 Here estimates are based on systematic analysis of two other extensive biographical dictionaries: Michele Rosi, *Dizionario del Risorgimento nazionale; Dalle origini a Roma capitale; Fatti e persone*, vols. 2–4: *Le persone* (Milan: Casa Dottor Francesco Vallardi, 1930–7); and Francesco Ercole, *Il Risorgimento italiano*, 5 vols., ser. 42, in A. Ribera (ed.), *Enciclopedia biografica e bibliografica "italiana"* (Milan: B.C. Tosi, 1936). For a fuller discussion of the use of biographical materials as sources for the study of migration, see Donna Gabaccia, "Class, Exile, and Nationalism at Home and Abroad: The Risorgimento," in Donna Gabaccia and Fraser Ottanelli (eds), *For Us There are No Frontiers* (forthcoming).

16 Rudolph J. Vecoli, "The Italian Diaspora, 1876–1976," in Robin Cohen (ed.), *The Cambridge Survey of World Migration* (Cambridge: Cambridge University Press, 1995), p. 114.

17 For a good overview, see Ercole Sori, *L'emigrazione italiana dall'unità alla seconda guerra mondiale* (Bologna: Il Mulino, 1979).

18 Silvia Pedraza, "Women and Migration," *Annual Review of Sociology* 17 (1991): 303–25.

19 Here I adopt Robert Harney's felicitous phrase for the male sojourners: "Men Without Women: Italian Migrants in Canada, 1885–1930," in Betty Boyd Caroli, Robert F. Harney, and Lydio F. Tomasi (eds), *The Italian Immigrant Woman in North America* (Toronto: Multicultural History Society of Ontario, 1978), pp. 79–101.

20 The term was developed by students of migration to Australia; see J.S. MacDonald, "Chain Migration, Ethnic Neighborhood Formation and Social Networks," *Millbank Memorial Fund Quarterly* 42 (January 1964): 82–91.

21 This method has been called "village–outward" by Samuel L. Baily, "The Future of Italian–American Studies: An Historian's Approach to Research in the Coming Decade," in Lydio Tomasi (ed.), *Italian Americans: New Perspectives in Italian Immigration and Ethnicity* (Staten Island: Center for Migration Studies, 1985), pp. 193–201. See also Samuel L. Baily, "The Village–Outward Approach to Italian Migration: A Case Study of Agnonesi Migration Abroad, 1885–1989," *Studi emigrazione* 29, 105 (March 1992): 43–68. Italian "village–outward" studies include Maria Rosaria Ostuni, *La diaspora politica dal Biellese* (Biellesi nel'mondo, Studi a cura di Valerio Castronovo), vol. 2 (Milan: Electa, 1995); Piero Ortoleva and Chiara Ottaviano (eds), *Sapere la strada: Percorsi e mestieri dei biellesi nel mondo* (Milan: Electa, 1986). Often, Italian scholars have preferred "region–outward" approaches: see, e.g.: Emilio Franzina, "Dopo il '76; Una regione all'estero," *Storia delle regioni*, vol. 2: *Il Veneto* (Turin: Einaudi, 1984); Giorgio Padoan (ed.), *Presenza, cultura, lingua e tradizioni dei veneti nel mondo* (Venice: Giunta Regionale, Regione Veneto, 1987).

22 Donna Gabaccia and Franca Iacovetta, "Women, Work, and Protest in the Italian Diaspora: Gendering Global Migration, Rethinking Family Economies, Nationalisms, and Labour Activism," *Labour/Le Travail* 42 (Fall 1998): 161–81.

23 *Atti della Giunta parlamentare per l'inchiesta agraria* (Rome: Forzani, 1881–6), vol. 11, pt. 2, p. 606.

24 *Annuario statistico italiano* (Rome: Commissariato dell'Emigrazione, 1927), Table VII.

25 Thus, even political scientists who abjure cultural interpretations make ungovernability a central theme of studies of Italian politics. See, e.g., Frederic Spotts and Theodor Weiser, *Italy — a Difficult Democracy* (Cambridge: Cambridge University Press, 1986).

26 Kerby A. Miller, *Emigrants and Exiles: Ireland and the Irish Exodus to North America* (New York: Oxford University Press, 1985); Matthew Frye Jacobson, *Special Sorrows; The Diasporic Imagination of Irish, Polish and Jewish Immigrants in the United States* (Cambridge, Mass.: Harvard University Press, 1995).

27 At best, case studies of Italians in particular countries are collected in a single anthology. Besides Pozzetta and Ramirez, *The Italian Diaspora*, and Bezza, *Gli italiani fuori d'Italia*, see Vanni Blengino, Emilio Franzina, and Adolfo Pepe, *La riscoperta delle Americhe; Lavoratori e sindacato nell'emigrazione italiana in America Latina 1870–1970* (Milan: Teti Editore, 1993); Emilio Franzina, *Gli italiani al nuovo mondo* (Milan: Arnoldo Mondadori Editore, 1995); Tomasi *et al.*, *The Columbus People*. The most explicitly comparative of such collections is Robert F. Harney and J. Vincenza Scarpaci (eds), *Little Italies in North America* (Toronto: Multicultural History Society of Ontario, 1981).

28 Although offering a forceful argument about the predominance of culture in determining global patterns of employment, Thomas Sowell, *Migrations and Cultures, A World View* (New York: HarperCollins, 1996), ch. 4 (on

Italians), rests on research that is too limited (and Anglocentric) to be trustworthy. Eighty years after its first publication, Foerster, *The Italian Emigration*, remains a better introduction.

29 But see Donna Gabaccia, "Italian History and gli italiani nel mondo, Parts I and II," *Journal of Modern Italian Studies* 2, 1 (1997): 45–66 and 3, 1 (1998): 73–97. See also the provocative book by Richard Bosworth, *Italy and the Wider World* (New York: Routledge, 1996), which offers two worthwhile chapters on emigration and Italian life worldwide.

30 Stephen Castles and Mark J. Miller, *The Age of Migration; International Population Movements in the Modern World* (New York: The Guilford Press, 1993), pp. 223–8.

31 Nina Glick Schiller, Linda Basch, and Cristina Blanc-Szanton (eds), *Towards a Transnational Perspective on Migration: Race, Class, Ethnicity, and Nationalism Reconsidered* (New York: Academic Press, 1992).

32 Arjun Appadurai, "Global Ethnoscapes: Notes and Queries for a Transnational Anthropology," in Richard Fox (ed.), *Recapturing Anthropology: Working in the Present* (Santa Fe, N.M.: School of American Research Press, 1991); and "Disjuncture and Difference in the Global Cultural Economy," *Public Culture* 2 (Spring 1990): 1–24.

33 Fernand Braudel, *The Mediterranean and the Mediterranean World in the Age of Philip II* (New York: Harper & Row, 1972).

34 Nina Glick Schiller, Linda Basch, and Cristina Blanc-Szanton, *Nations Unbound: Transnational Projects, Postcolonial Predicaments, and Deterritorialized Nation-States* (New York: Gordon & Breach, 1994).

35 Here I borrow the terminology, but not the analytical point of view, of John A. Armstrong, "Mobilized and Proletarian Diasporas," *American Political Science Review* 70 (1976): 393–408.

1 BEFORE ITALIANS: MAKING ITALIAN CULTURE AT HOME AND ABROAD

1 Luciana Cocito (ed.), *Le rime volgari dell'Anonimo Genovese* (Genoa: M. Bozzi, 1966).

2 I am aware that I am not using the term nation as many English speakers do, especially in North America. Americans commonly distinguish nations from peoples, defining nations through their loyalties to states, governments, and political ideologies, and peoples through cultural, historical, and linguistic solidarities. Given the nature of the U.S. — a nation defined by particular political-economic beliefs, not a common cultural heritage — this is understandable. Following Benedict Anderson, I instead see a nation as an imagined community, that is as a people. A nation may, or may not, desire a state, or government, to unite or represent it, however. The relations among nations, states, and nationalism are complex and hotly debated. In order to avoid confusion I use the dialect term "natio" or "nation" for the urban and regional communities that existed in Italy prior to 1861. I reserve nation (without quotation marks) for the modern group that created the Italian national state in 1861–70. Readers should bear in mind, however, that nationalism — the idea that each nation needs its own state, and the political movement that accompanies

that idea — itself asserted the right to define the nation in the modern period. Everywhere in Europe, nationalists were a small subset of the peoples who became citizens of nation states. States, in turn, often worked hard to encourage nationalism, to create and expand a sense of belonging in the nations that justified their existence. English speakers will find good introductions to these issues in: Benedict Anderson, *Imagined Communities: Reflections on the Origin and Spread of Nationalism* (New York: Verso, 1983); Anthony D. Smith, *The Ethnic Origins of Nations* (Oxford: Oxford University Press, 1986); John A. Armstrong, *Nations before Nationalism* (Chapel Hill: University of North Carolina Press, 1981); Miroslav Hroch, *Social Preconditions of National Revival in Europe: A Comparative Analysis of the Social Composition of Patriotic Groups among the Smaller European Nations* (Cambridge: Cambridge University Press, 1985); Étienne Balibar, "The Nation Form: History and Ideology," *Review of the Fernand Braudel Center* 13, 3 (1990): 329–61.

3 Although my purpose is to problematize the use of the terms "Italy" and "Italian" before the creation of a nation and national state of Italians and Italy, I have chosen not to disfigure the text with the quotation marks with which I can distance myself from their everyday meaning. Similarly I have chosen to use Italian terms, without italicization, when no exact equivalent exists in English.

4 Giuseppe Galasso, *Storia d'Italia*, vol. 1: *L'Italia come problema storiografico* (Turin: Unione Tipografico Ed. Torinese, 1979), ch. 4.

5 Charles Tilly, *Coercion, Capital and European States, AD 990–1990* (Oxford: Basil Blackwell, 1990), pp. 45–7.

6 Jacques Le Goff, "L'Italia fuori d'Italia: L'Italia nello specchio del Medioevo," pt. I, in Ruggiero Romano and Corrado Vivanti (eds), *Storia d'Italia*, vol. 2: *Dalla caduta dell'Impero romano al secolo XVIII*, part 2 (Turin: G. Einaudi, 1974), p. 1943.

7 Maria S. Haynes, *The Italian Renaissance and its Influence on Western Civilization* (Lanham, Md.: University Press of America, 1991).

8 Le Goff, "L'Italia nello specchio del Medioevo," p. 2057; Fernand Braudel, "L'Italia fuori d'Italia: Due secoli e tre Italie," pt. 2, in Romano and Vivanti, *Dalla caduta dell'Impero romano al secolo XVIII*, part 2, p. 2091.

9 Quoted in Eric Wolf, *Europe and the People Without History* (Berkeley: University of California Press, 1982), p. 71.

10 Janet Abu-Lughod, *Before European Hegemony, The World System, A.D. 1250–1350* (New York: Oxford University Press, 1989).

11 I borrow this image from William H. McNeill, *Venice, the Hinge of Europe 1081–1797* (Chicago, Ill.: University of Chicago Press, 1974).

12 Italy's early nationalist historians seem to have first popularized the phrase. See Vincenzo Gioberti, *Del primato morale e civile degli italiani* (Capolago: Elvetica, 1846). For a sense of the power of this primacy see Indro Montanelli and Roberto Gervaso, *L'Italia dei secoli d'oro; Il Medio Evo dal 1250 al 1492* (Milan: Rizzoli, 1967).

13 Carlo Cipolla, "The Decline of Italy: The Case of a Fully Matured Economy," *Economic History Review* 2nd series V, 2 (1952): 178; see also Paolo Malanima, *La fine del primato: Crisi e riconversione nell'Italia del seicento* (Milan: B. Mondadori, 1998).

14 Immanuel Wallerstein, *The Modern World System II: Mercantilism and the*

Consolidation of the European World-Economy, 1600–1750 (New York: Academic Press, 1980).

15 Gino Luzzatto, *An Economic History of Italy; From the Fall of the Roman Empire to the Beginnings of the Sixteenth Century* (New York: Barnes & Noble, 1961); Harry A. Miskimin, *The Economy of Early Renaissance Europe, 1300–1460* (Englewood Cliffs, N.J.: Prentice-Hall, 1969).

16 Daniel Philip Waley, *The Italian City-Republics* (New York: McGraw Hill, 1973).

17 Bernard Doumerc, "L'immigration à Venise à la fin du Moyen-Age," in Simonetta Cavaciocchi (ed.), *Le migrazioni in Europea, secc. XIII–XVIII* (Florence: Le Monnier, 1994); Alain Ducellier, "Marché du travail, esclavage et travailleurs immigrés dans le nord-est de l'Italie (fin du XIVe siècle–milieu du XVe siècle)," in Michel Balard (ed.), *État et colonisation au Moyen Age* (Lyons: La Manufacture, 1989), pp. 217–49; Carmelo Trasselli, "Genovesi in Sicilia," *Atti della Società Ligure di Storia Patria* n.s. 9 (1969): 153–78.

18 Aleksandrovna Kotelnikova Liubov, *Mondo contadino e città in Italia dall'XI al XIV secolo; Dalle fonti dell'Italia centrale e settentrionale* (Bologna: Il Mulino, 1982). See also Trevor Dean and Chris Wickham, *City and Countryside in Late Medieval and Renaissance Italy: Essays Presented to Philip Jones* (London: Ronceverte, 1990).

19 Giulio Bollati, "L'italiano," in Romano and Vivanti, *Storia d'Italia*, vol. 1, pp. 951–67, passim.

20 Lauro Martines (ed.), *Violence and Disorder in Italian Cities, 1200–1500* (Berkeley: University of California Press, 1972).

21 Jacob Burckhardt, *The Civilization of the Renaissance in Italy* (Oxford: Phaidon Press, 1945), part I; see also Waley, *The Italian City-Republics*, ch. 5.

22 Quoted in Maurizio Viroli, *For Love of Country: An Essay on Patriotism and Nationalism* (Oxford: Clarendon Press, 1997), p. 27.

23 Charles Verlinden, "From the Mediterranean to the Atlantic: Aspects of an Economic Shift (12th–18th Century)," *Journal of European Economic History* 1, 3 (Winter 1972): 625–46; Ruth Pike, *Enterprise and Adventure: The Genoese in Seville and the Opening of the New World* (Ithaca, N.Y.: Cornell University Press, 1966); Alan K. Smith, *Creating a World Economy: Merchant Capital, Colonialism, and World Trade, 1400–1825* (Boulder, Colo.: Westview Press, 1991).

24 Le Goff calls Italy in these years, "più desiderabile che mai, più debole che mai," in "L'Italia nello specchio del Medioevo," p. 2085.

25 Fernand Braudel, "A Model for the Analysis of the Decline of Italy," *Review of the Fernand Braudel Center* 2, 4 (Spring 1979): 662.

26 Galasso, *L'Italia come problema storiografico*, p. 81.

27 Peter Burke, *The Italian Renaissance: Culture and Society in Italy* (Princeton, N.J.: Princeton University Press, 1986).

28 Quoted in H.G. Koenigsberger, *The Practice of Empire* (Ithaca, N.Y.: Cornell University Press, 1969), p. 48.

29 Le Goff, "L'Italia nello specchio del Medioevo," p. 1994. See e.g. Johann Wolfgang von Goethe, *Italienische Reise* (Munich: Wilhelm Goldmann Verlag, 1961). "Wo die Zitronen blühn" is from Goethe's *Wilhelm Meisters Lehrjahre*.

30 Robert Foerster, *The Italian Emigration of Our Times* (New York: Russell & Russell, 1968), p. 5.

31 Giovanni Pizzorusso and Matteo Sanfilippo, "Rassegna storiografica sui fenomeni migratori a lungo raggio in Italia dal basso Medioevo al secondo dopoguerra," *Bollettino di demografia storica* 13 (1990).

32 Antonio Gramsci, *Prison Notebooks*, ed. and trans. Joseph A. Buttigieg (New York: Columbia University Press, 1996), vol. 2, pp. 102–4, 260–1.

33 See Leo Benvenuti, *Dizionario degli italiani all'estero* (Florence, 1890), and Francesco Fortunato Carloni, *Gl'italiani all'estero dal secolo VII ai di nostri*, 3 vols. (Città di Castello: Tip. Lapi, 1888–1908). Carloni's project, the most ambitious, surveyed the arts, sciences, and religious and secular administration.

34 Born in 1889, Imperatori was a filiopietist and nationalist whose efforts to memorialize Italy's great migrants began under fascism, but continued into the postwar years. See his early works: Ugo E. Imperatori, *Italia prodiga (gli italiani all'estero)* (Milan: Alpes, 1925), and Ugo E. Imperatori, *Italiani fuori d'Italia* (Milan: Oberdan Zucchi, 1937).

35 Ugo E. Imperatori, *Dizionario di italiani all'estero* (Genoa: L'Emigrante, 1956).

36 Jacques Dupaquier, "Macro-migrations en Europe (XVI–XVIIIe siècles," in Cavaciocchi, *Le migrazioni*, pp. 70–1; see also George Gush, *Renaissance Armies, 1480–1650* (Cambridge: Patrick Stephens, 1975), p. 19; Antonio Ivan Pini, *Le grandi migrazioni umane nell'antichità e nel Medioevo* (Florence: La Nuova Italia, 1969).

37 Fernand Braudel, *The Mediterranean and the Mediterranean World in the Age of Philip II* (New York: Harper & Row, 1973), vol. 2, pp. 771, 776–7.

38 Italian scholars, intellectuals, poets, and writers also remained a persistent, and large, group of migrants over the centuries. Other important migratory producers of civiltà italiana included the employees of foreign dynasties (diplomats and government servants were 2 to 7 percent), military men (4 to 8 percent), and lawyers specializing in Catholic or Roman law.

39 There is a rich literature on the trading empires of Genoa and Venice. See Georges Jehel, *Les Genois en Mediterranée occidentale (fin Xe–début XIVe siècle): Ébauche d'une stratégie pour un empire* (Amiens: Centre d'Histoire des Sociétés, Université de Picardie, 1993); Geo Pistarino, *Genovesi d'Oriente* (Genoa: Civico Istituto Colombiano, 1990); David Sanderson Chambers, *The Imperial Age of Venice, 1380–1580* (London: Thames & Hudson, 1970); John Julius Cooper Norwich, *Venice: The Rise to Empire* (London: Allen Lane, 1977).

40 David Jacoby, "The Migration of Merchants and Craftsmen: A Mediterranean Perspective (12th–15th Century)," in Cavaciocchi, *Le migrazioni*, pp. 533–60.

41 Patrizia Audenino, *Un mestiere per partire: tradizione migratoria, lavoro, e comunità in una vallata alpina* (Milan: Franco Angeli, 1990).

42 Between 3 and 9 percent of all migrants before 1500, southerners were 12 percent of migrants in the 1500s, and over 20 percent in the years between 1790 and Italian unification.

43 A student of Italy's earliest merchant migrants, Sapori argued that they "seemed, more or less unconsciously, to pave the way for the Risorgi-

mento" — that is, for Italian nationalism and its belief in a unified Italian people. Armando Sapori, *The Italian Merchant in the Middle Ages* (New York: W.W. Norton, 1970), pp. 14–15.

44 Raymond De Roover, *La Communauté des marchands lucquois à Bruges de 1377 à 1404*, quoted in Abu-Lughod, *Before European Hegemony*, p. 90; see also Frédéric Mauro, "Merchant Communities, 1350–1750," in James D. Tracy (ed.), *The Rise of Merchant Empires; Long-Distance Trade in the Early Modern World, 1350–1750* (Cambridge: Cambridge University Press, 1990), pp. 255–86.

45 E. Ashtor, "The Venetian Supremacy in Levantine Trade: Monopoly or Pre-Colonialism?," *Journal of European Economic History* 3 (1974): 5–53; Robert S. Lopez, "Market Expansion: The Case of Genoa," *Journal of Economic History* 24 (1964): 445–64; Robert S. Lopez, *Storia delle colonie genovesi nel Mediterraneo* (Genoa: Marietti, 1996).

46 McNeill, *Venice*, p. 4. See H. Mackenzie, "The Anti-Foreign Movement in England 1231–1232," cited in Le Goff, "L'Italia nello specchio del Medioevo," p. 2001n. On French opposition, see J.F. Dubost, "Les Italiens en France aux XVIe et XVIIe siècles (1570–1600)," History thesis, University of Paris I, 1992.

47 Le Goff, "L'Italia nello specchio Medioevo," pt. 3.

48 Michele Cassandro, "I forestieri a Lione nel '400 e '500: La nazione fiorentina," in Gabriella Rossetti (ed.), *Dentro la città, Stranieri e realtà urbane nell'Europa dei secoli XII–XVI* (Naples: GISEM-Liguori, 1989), pp. 151–62.

49 As speakers of a vernacular offshot of Latin, Italians and French together formed this nation, separate from the English and German nation of speakers of non-Romance languages. At the university in Padua after 1260, by contrast, the nations of students included ultramontani ("beyond the mountains" — e.g. Lombards, Tuscans, Romans) and citramontani (Le Goff, "L'Italia nello specchio del Medioevo," p. 1952).

50 Wolf, *Europe and the People Without History*, p. 71.

51 Sapori, *The Italian Merchant*, pp. 14–15.

52 Le Goff, "L'Italia nello specchio del Medioevo," p. 1952.

53 Giuliano Pinto, "La politica demografica delle città," in Rinaldo Comba (ed.), *Strutture familiari, epidemie, migrazioni nell'Italia medioevale* (Naples: ESI, 1984), pp. 19–43; W.W. Bowsky, "Medieval Citizenship: The Individual and the State in the Commune of Siena, 1287–1355," *Studies in Medieval and Renaissance History* 4 (1967): 193–243; Elena Guarini Fasano, "La politica demografica delle città italiane nell'età moderna," in *Società italiana di demografia storica; La demografia storica delle città italiane* (Bologna: Clueb, 1982), pp. 149–91.

54 Giuseppe Marco Antonio Baretti, quoted in Alfredo Comandini, *L'Italia nei cento anni del secolo XIX (1801–1900); Giorno per giorno illustrata*, vol. 1: *1801–1825* (Milan: Antonio Vallardi, 1900–1), p. viii.

55 In modern Italian little is "piccolo." See maps in K. Jaberg and J. Jud, *Sprach- und Sachatlas Italiens und der Südschweiz* (Zofingen: Ringier, 1933), cited in Galasso, *L'Italia come problema storiografico*, p. 180.

56 Baretti, quoted in Comandini, *L'Italia.*

2 MAKING ITALIANS AT HOME AND ABROAD, 1790–1893

1 A peasant of Lombardy, 1878, quoted in Jerre Mangione and Ben Morreale, *La Storia; Five Centuries of the Italian American Experience* (New York: HarperCollins, 1992), p. 33.
2 *Atti della Giunta parlamentare per l'inchiesta agraria* (Rome: Forzani, 1881–6), vol. 4, fascicolo (hereafter fasc.) 1, p. 17.
3 English speakers will want to start with Stuart J. Woolf, *The Italian Risorgimento* (London: Longman, 1969); Derek Beales, *The Risorgimento and the Unification of Italy* (London: Allen & Unwin, 1971); Stuart J. Woolf, *A History of Italy, 1700–1860; The Social Constraints of Political Change* (London: Methuen, 1979); Lucy Riall, *The Italian Risorgimento: State, Society and National Unification* (London: Routledge, 1994). In Italian, start with G. Sabbatucci and V. Vidotto (eds), *Storia d'Italia*, vol. 1: *Le premesse dell'unità* (Roma-Bari: Laterza, 1995). Also useful is a recent historiographical assessment, Rosario Romeo, "Il Risorgimento nel dibattito contemporaneo," *Rassegna storica del Risorgimento* 85, 1 (January–March 1998): 3–16.
4 Zeffiro Ciuffoletti, "L'esilio nel Risorgimento," in Maurizio Degl'Innocenti (ed.), *L'esilio nella storia del movimento operaio e l'emigrazione economica* (Manduria: Piero Lacaita, 1992), pp. 53–60. See the biographies in Leone Carpi, *Il Risorgimento italiano: biografie illustri italiane*, 4 vols. (Milan: Dottor Francesco Vallardi, 1888). References to the characteristics and experiences of nationalists as exiles are based on my analysis of over 1900 biographies of nationalist activists described in Michele Rosi, *Dizionario del Risorgimento nazionale; Dalle origini a Roma capitale; Fatti e persone* (Milan: Casa Dottor Francesco Vallardi, 1930–7), and Francesco Ercole, *Il Risorgimento italiano*, 5 vols., ser. 42, in A. Ribera (ed.), *Enciclopedia biografica e bibliografica "italiana"* (Milan: B.C. Tosi, 1936).
5 On diaspora nationalism, see Ernest Gellner, *Nations and Nationalism* (Ithaca, N.Y.: Cornell University Press, 1983), pp. 102–3. On the Italian case, see Matteo Sanfilippo, "Nationalisme, 'italianité' et émigration aux Amériques (1830–1990)," *European Review of History* 2 (1995): 177–91.
6 Jacques Godechot, "Risorgimento et régionalisme en Italie," *Risorgimento* 2 (1981): 63–79
7 Ugo Foscolo, "Ultime lettere di Iacopo Ortis," *Opere di Ugo Foscolo*, ed. Mario Puppo (Milan: Ugo Mursia, 1973), p. 473.
8 Much of the work on the early years of the Risorgimento is now dated, and a product of the nationalist and fascist movements of the late nineteenth and early twentieth centuries. Still useful is Alberto Maria Ghisalberti, *Gli albori del Risorgimento italiano (1700–1815)* (Rome: P. Cremonese, 1931); see also Umberto Marcelli, *Riforme e rivoluzione in Italia nel secolo XVIII* (Bologna: R. Patron, 1964).
9 See the useful articles collected in John A. Davis and Paul Ginsborg (eds), *Society and Politics in the Age of the Risorgimento: Essays in Honour of Denis Mack Smith* (New York: Cambridge University Press, 1991), especially the articles by John Davis, "1799: The Santafede and the Crisis of the Ancien Regime in Southern Italy," and Franco Della Peruta, "War and Society in Napoleonic Italy."

10 Guido Formigoni, *L'Italia dei cattolici: Fede e nazione dal Risorgimento alla Repubblica* (Bologna: Il Mulino, 1998), ch. 1.
11 See entries in Rosi, *Dizionario del Risorgimento*; Ercole, *Il Risorgimento italiano*, vol. 2; Bianca Montale, *L'emigrazione politica in Genova ed in Liguria, 1849–1859* (Savona: Sabatelli, 1982).
12 Gian Biagio Furiozzi, *L'emigrazione politica in Piemonte nel decennio preunitario* (Florence: Leo S. Olschki Editore, 1979), p. 12.
13 Here, I extend to the Italian biographical dictionaries the method utilized by Miroslav Hroch, *Social Preconditions of National Revival in Europe: A Comparative Analysis of the Social Composition of Patriotic Groups among the Smaller European Nations* (Cambridge: Cambridge University Press, 1985).
14 Roberto Romani, *L'economia politica del Risorgimento italiano* (Turin: Bollati Boringhieri, 1994).
15 Adrian Lyttelton, "The Middle Classes in Liberal Italy," in Davis and Ginsborg, *Society and Politics in the Age of the Risorgimento*.
16 Entries for Scardi and De Thomasis in Rosi, *Dizionario del Risorgimento*. See also Laura Pisano and Christiane Vauvy, *Parole inascoltate; Le donne e la costruzione dello stato-nazione in Italia e in Francia 1789–1860* (Rome: Riuniti, 1994); Antonio Spinosa, *Italiane; Il lato segreto del Risorgimento* (Milan: Mondadori, 1994); Daniela Pizzagalli, *L'amica: Carla Maffei e il suo salotto nel Risorgimento italiano* (Milan: Mondadori, 1997).
17 Pietro Brunello, *Ribelli, questuanti e banditi; Proteste contadine in Veneto e in Friuli, 1814–1866* (Venice: Marsilio, 1981); Sandro Lombardini, *Rivolte contadine in Europa (secoli XVI–XVIII)* (Turin: Loescher, 1983); Oscar DeSimplico, *Le rivolte contadine in Europa* (Rome: Riuniti, 1986).
18 Besides John A. Davis, *Conflict and Control: Law and Order in Nineteenth-Century Italy* (Houndmills, Basingstoke: Macmillan Education, 1988), see Steven C. Hughes, *Crime, Disorder, and the Risorgimento: The Politics of Policing in Bologna* (Cambridge: Cambridge University Press, 1994); David I. Kertzer, *Sacrificed for Honor: Italian Infant Abandonment and the Politics of Reproductive Control* (Boston, Mass.: Beacon Press, 1993).
19 Eric J. Hobsbawm, *Primitive Rebels: Studies in Archaic Forms of Social Movement in the 19th and 20th Centuries* (New York: W.W. Norton, 1959).
20 Quoted in Francesco Leoni, *Storia della contrarivoluzione in Italia (1789–1859)* (Naples: Guida, 1974), p. 44.
21 See entries on Mazziotti in Ercole, *Il Risorgimento italiano*, vol. 2, and Rosi, *Dizionario del Risorgimento*; Bianca Marcolongo, *Le origini della Carboneria e le società segrete nell'Italia meridionale dal 1810 al 1820* (Sala Bolognese: A. Forni, 1983). On the democratic nationalists more generally see Giuseppe Berti, *I democratici e l'iniziativa meridionale nel Risorgimento* (Milan: Feltrinelli, 1962); Clara Lovett, *The Democratic Movement in Italy, 1830–1876* (Cambridge, Mass.: Harvard University Press, 1982); and Franco della Peruta, *I democratici e la rivoluzione italiana (Dibattiti ideali e contrasti politici all'indomani del 1848)* (Milan: Feltrinelli, 1974).
22 Nadia Urbinati, "'A Common Law of Nations': Giuseppe Mazzini's Democratic Nationality," *Journal of Modern Italian Studies* 1, 2 (1996): 197–222. Recent general studies of Mazzini include Roland Sarti, *Mazzini, A Life for the Religion of Politics* (Westport, Conn.: Greenwood Publishing Group, 1997); Denis Mack Smith, *Mazzini* (New Haven, Conn.: Yale University Press, 1994); Luigi Ambrosoli, *Giuseppe Mazzini, Una vita per l'unità*

d'Italia (Manduria: Piero Lacaita, 1993); Romano Bracalini, *Mazzini: Il sogno dell'Italia onesta* (Milan: A. Mondadori, 1994).

23 Spencer Di Scala, *Italy: From Revolution to Republic, 1700 to the Present* (Boulder, Colo.: Westview Press, 1995), pp. 66–7. See V. Douglas Scotti, "La guerriglia negli scrittori risorgimentali italiani primo e dopo il 1848–1849," *Il Risorgimento* 27 (1976): 106–10.

24 Giuseppe Mazzini, *L'amore e la missione della donna, Pensieri* (Genoa: Libreria Moderna, 1921); see also Judith Jeffrey Howard, "Patriot Mothers in the Post-Risorgimento: Women after the Italian Revolution," in Carol R. Berkin and Clara M. Lovett (eds), *Women, War, and Revolution* (New York: Holmes & Meier, 1980).

25 See entries on Pisacane, Vigano, and Salvi in Rosi, *Dizionario del Risorgimento*, in Ercole, *Il Risorgimento italiano*, vol. 3, and in Francesco Ercole, *I martiri* (Rome: Tosi, 1946). See also Luigi Bulferetti, *Socialismo risorgimentale* (Turin: G. Einaudi, 1949). Conflicting interpretations of Garibaldi as socialist are in Richard Hostetter, *The Italian Socialist Movement*, vol. 1: *Origins (1860–1872)* (Princeton, N.J.: D. Van Nostrand, 1958); Letterio Briguglio, *Garibaldi e il socialismo* (Milan: SugarCo, 1982).

26 M.A. Fonzi Columba, "L'emigrazione," in *Bibliografia dell'età del Risorgimento in onore di A.M. Ghisalberti*, vol. 2 (Florence: Olschki, 1972); Alessandro Galante Garrone, "L'emigrazione politica italiana del Risorgimento," *Rassegna storica del Risorgimento* 41 (1954): 229–42; and the special issue "Gli italiani nel mondo ed il Risorgimento," *Il veltro* 5–6 (1961).

27 Compared to the migrations that followed Italian unification, the migrations of these years have received little study. Useful are: John E. Zucchi, *The Little Slaves of the Harp: Italian Child Street Musicians in Nineteenth-Century Paris, London, and New York* (Montreal: McGill-Queen's University Press, 1992); Francesco Surdich, "I viaggi, i commerci, le colonie: Radici locali dell'iniziativa espansionistica," in Antonio Gibelli and Paride Rugafiori (eds), *La Liguria* (Turin: G. Einaudi, 1994), pp. 457–509. For the Americas, Chiara Vangelista, "Traders and Workers: Sardinian Subjects in Argentina and Brazil," in George E. Pozzetta and Bruno Ramirez (eds), *The Italian Diaspora: Migration across the Globe* (Toronto: Multicultural History Society of Ontario, 1992), pp. 37–50; Emilio Franzina, *Gli italiani al nuovo mondo* (Milan: Arnoldo Mondadori Editore, 1995), chs. 1 and 2; José C. Chiaramonte, "Notas sobre la presencia italiana en el litoral argentino en la primera mitad del siglo XIX," in Fernando J. Devoto and Gianfausto Rosoli, *L'Italia nella società Argentina* (Rome: Centro Studi Emigrazione, 1988), pp. 44–58.

28 Walter W. Willcox and Imre Ferenczi, *International Migrations* (New York: National Bureau of Economic Research, 1929), vol. 1, U.S., Table IX and Argentina, Table V.

29 Leone Carpi, *Delle colonie e dell'emigrazione d'italiani all'estero sotto l'aspetto dell'industria, commercio ed agricoltura* (Milan: Lombarda, già D. Salvi, 1874), vols. 1–2.

30 My estimates, based on biographies in Imperatori, *Dizionario di italiani all'estero*.

31 Franzina, *Gli italiani al nuovo mondo*, p. 15.

32 See entries on Bosso and Ghesa in Rosi, *Dizionario del Risorgimento*; on padroni, see Chapter 3 of this book. On journalists, see Leonidi Ballestreri,

"Patrioti del Risorgimento nella storia del giornalismo di paesi stranieri," *Miscellanea di storia ligure in onore di Giorgio Falco* (Milan: Feltrinelli, 1962), pp. 411–20.

33 Fernando J. Devoto, "Liguri nell'America australe: Reti sociali, immagini, identità," in Gibelli and Rugafiori, *La Liguria*, 653–88; see also Fernando J. Devoto, "Inventing the Italians? Images of Immigrants in Buenos Aires, 1810–1880," in Pozzetta and Ramirez, *The Italian Diaspora*, pp. 62–85.

34 Zucchi, *The Little Slaves of the Harp*, ch. 3; Lucio Sponza, *Italian Immigrants in Nineteenth-Century Britain: Realities and Images* (Leicester: Leicester University Press, 1988).

35 See entries for all in Rosi, *Dizionario del Risorgimento*; Ercole, *Il Risorgimento italiano*. For background, Franco Valsecchi, "Il sistema costituzionale inglese nel pensiero politico risorgimentale," *Rassegna storica del Risorgimento* 46 (January–March 1979): 25–37; Emilia Morelli, "Gli esuli italiani e la società inglese nella prima metà dell'Ottocento," *Rassegna storica del Risorgimento* 46 (January–March 1979): 3–13; Margaret Campbell Walker Wicks, *The Italian Exiles in London, 1816–1848* (Manchester: Manchester University Press, 1937).

36 See entry for Rolandi in Rosi, *Dizionario del Risorgimento*. On Mazzini's London years, see Mack Smith, *Mazzini*, pp. 37–9.

37 Bianca Montale, *Mazzini e le origini del movimento operaio italiano; Appunti di storia del Risorgimento* (Genoa: Tilgher, 1976); Francesco Fiumara, *Mazzini tra le brume di Londra: Unione operai italiani; Apostolato popolare* (Reggio Calabria: La Procellaria, 1987).

38 Ema Cibotti, "Mutualismo y política en un estudio de caso; La sociedad 'Unione e Benevolenza' en Buenos Aires entre 1858 y 1865," in Devoto and Rosoli, *L'Italia nella società Argentina*, pp. 241–64. See also entry for Basilio Cittadini in Imperatori, *Dizionario di italiani all'estero*; Samuel L. Baily, "Las sociedades de ayuda mutua y el desarrollo de una comunidad italiana en Buenos Aires, 1858–1918," *Desarrollo economica* 21 (1982): 488–9; Fernando J. Devoto, "La primera elite politica italiana de Buenos Aires (1852–1880)," *Studi emigrazione* 26, 94 (1989): 168–93.

39 Zucchi, *The Little Slaves of the Harp*, ch. 2; John Zucchi, "'Les petits Italiens': Italian child street musicians in Paris, 1815–1875," *Studi emigrazione* 97 (1990): 27–54.

40 See entries for Cannonieri, Dragonetti, Montazio, and Belgioso in Rosi, *Dizionario del Risorgimento*.

41 Luciano Iorizzo and Salvatore Mondello, *The Italian Americans*, rev. edn. (Boston, Mass.: Twayne, 1980), p. 26; Salvatore Candido, "L'unione mazziniana nelle Americhe e la Congrega di New York della 'Giovine Italia' (1842–1852)," *Bollettino della Domus Mazziniana* 17 (1972): 123–75.

42 See entries in Rosi, *Dizionario del Risorgimento*; Ercole, *Il Risorgimento italiano*, vol. 3, and vol. 5, *I combattenti*.

43 Guiseppe Ferrari, *Filosofia della rivoluzione* (Milan: Mazorati, 1970). See entries on Ferrari in Rosi, *Dizionario del Risorgimento*, Ercole, *Il Risorgimento italiano*, vol. 2, and Franco Andreucci and Tommaso Detti (eds), *Il movimento operaio italiano, Dizionario biografico, 1853–1943* (Rome: Riuniti, 1975–8). See also Clara M. Lovett, *Giuseppe Ferrari and the Italian Revolution* (Chapel Hill: University of North Carolina Press, 1979).

44 Surprisingly, the Parisian circle of exiles has not received the attention it warrants. A starting place is Salvatore Carbone, *Fonti per la storia del Risorgimento italiano negli archivi nazionali di Parigi. I rifugiati italiani 1815–1830* (Rome: Istituto per la Storia del Risorgimento Italiano, 1962).
45 See entries in Rosi, *Dizionario del Risorgimento*, and Ercole, *I martiri*.
46 An introduction to the vast bibliography on Garibaldi is Anthony P. Campanella, *Giuseppe Garibaldi e la tradizione garibaldina; Una bibliografia dal 1807 al 1970*, 2 vols. (Geneva: Comitato dell'Istituto Internazionale di Studi Garibaldina, 1970). On the early garibaldini, see Carlo Jean, "Garibaldi e il volontariato italiano nel Risorgimento," *Rassegna storica del Risorgimento* 69, 4 (1982): 399–419.
47 See Franzina, *Gli italiani al nuovo mondo*, pp. 102–12.
48 Elsa Feraboli, "Il primo esilio di Garibaldi in America, 1833–1848," *Rassegna storica del Risorgimento* 19, 2 (1932): 247–82; Salvatore Candido, *Giuseppe Garibaldi nel Rio della Plata, 1841–1848* (Florence: Valmartina, 1972); Ivan Boris, *Gli anni di Garibaldi in Sud America, 1836–1848* (Milan: Longanesi, 1970).
49 See entries in Rosi, *Dizionario del Risorgimento*, and Ercole, *Risorgimento italiano*, vol. 3.
50 Assessments of Garibaldi's life from a variety of perspectives include: Denis Mack Smith, *Garibaldi, A Great Life in Brief* (New York: Knopf, 1956); Romano Ugolini, *Garibaldi: Genesi di un mito* (Rome: Ed. Dell'Ateneo, 1982); Indro Montanelli, *Garibaldi* (Milan: Rizzoli, 1982).
51 Denis Mack Smith, *Cavour and Garibaldi, 1860: A Study in Political Conflict* (Cambridge: Cambridge University Press, 1954).
52 Ugo Zaniboni Ferino, *Bezzecca 1866: La campagna garibaldina fra l'Adda e il Garda* (Trento: Museo Trentino del Risorgimento e della Lotta per la Libertà, 1987); Robert Molis, *Les Francs-tireurs et les Garibaldi: Soldats de la République: 1870–1871 en Bourgogne* (Paris: Tiresias, 1995).
53 A mere sixty-seven were elected to Italy's parliament or served as senators (twenty-one as conservatives and forty-three as republican or other critics of the moderates); another ten — including the prominent Carlo Cattaneo — declined election or government appointments, disdaining service under monarchy.
54 *Edizione nazionale degli scritti di Giuseppe Garibaldi* (Rome: Istituto per la Storia del Risorgimento Italiano, 1973–), vol. 6, p. 306.
55 Francesco Crispi, *Carteggi politici inediti di Francesco Crispi (1860–1900)*, ed. T. Palamenghi-Crispi (Rome: Universelle, 1912), p. 457. See also Christopher Duggan, "Francesco Crispi, 'Political education' and the Problem of Italian National Consciousness, 1860–1896," *Journal of Modern Italian Studies* 2, 2 (1997): 141–66.
56 Zeffiro Ciuffoletti, "Stato senza nazione," in *La costruzione dello stato in Italia e Germania* (Manduria: Lacaita, 1993).
57 Alberto Caracciolo, *L'inchiesta agraria Jacini* (Turin: G. Einaudi, 1973).
58 For historians' assessments of the period following unification, see Denis Mack Smith, *Italy, A Modern History* (Ann Arbor: University of Michigan Press, 1969); Di Scala, *Italy: From Revolution to Republic*; Nino Valeri (ed.), *Storia d'Italia* (Turin: Unione, 1965); Giorgio Candeloro, *Storia d'Italia moderna*, vols. 5–6 (Milan: Feltrinelli, 1968, 1970).

59 Gianni Toniolo, *Economic History of Liberal Italy, 1850–1918* (London: Routledge, 1990); Vera Zamagni, *The Economic History of Liberal Italy, 1860–1990* (New York: Oxford University Press, 1993).

60 From a rich literature, see Arturo Carlo Jemolo, *Church and State in Italy, 1850–1950* (Oxford: Blackwell, 1960); Giovanni Spadolini, *Opposizione cattolica da Porta Pia al '98* (Florence: Vallecchi, 1961); Francesco Traniello, *Città dell'uomo: Cattolici, partito e stato nella storia d'Italia* (Bologna: Il Mulino, 1990).

61 Quoted in John Dickie, "Stereotypes of the Italian South 1860–1900," in Robert Lumley and Jonathan Morris (eds), *The New History of the Italian South; The Mezzogiorno Revisited* (Exeter: University of Exeter Press, 1997), p. 122.

62 The quote is from A. Niceforo, *L'Italia barbara contemporanea (Studi ed appunti)* (Milan-Palermo: Remo Sandron, 1989), p. 6 in Gabriella Gribaudi, "Images of the South: The Mezzogiorno as seen by Insiders and Outsiders," in Lumley and Morris, *The New History of the Italian South*, p. 95.

63 Quoted in Davis, *Conflict and Control*, p. 180.

64 Francesco Gaudioso, *Calabria ribelle: Brigantaggio e sistemi repressivi nel Cosentino (1860–1870)* (Milan: Franco Angeli, 1987); Gaetano Cingari, *Brigantaggio, proprietari e contadini nel sud (1799–1900)* (Reggio Calabria: Meridionali Riuniti, 1976); Roberto Martucci, *L'emergenza e tutela dell'ordine pubblico nell'Italia liberale: Regime eccezionale e leggi da repressione dei reati di brigantaggio (1861–1865)* (Bologna: Il Mulino, 1980); Stefano Camelli, *Al suono delle campane: Indagine su una rivolta contadino: I moti del macinato (1869)* (Milan: Franco Angeli, 1984).

65 Quoted in Davis, *Conflict and Control*, p. 53.

66 Franco Lo Piparo, "Sicilia linguistica," and "L'italianizzazione linguistica di massa," in Maurice Aymard and Giuseppe Giarrizzo (eds), *Storia d'Italia: Le regione dall'unità a oggi*, vol. 6: *La Sicilia* (Turin: G. Einaudi, 1987), p. 790. On the more general issue of linguistic difference, see Tullio De Mauro, *Storia linguistica dell'Italia unita* (Bari: Laterza, 1963).

67 On the Risorgimento nationalists' faith in education, see Giuseppe Mazzini, *Educazione e democrazia: Antologia di scritti di Mazzini*, ed. Antonio Bandini Buti (Milan: Cisalpino-Goliardica, 1972). Useful general studies include Giuseppe Natale and Francesco Paolo Colucci, *La scuola in Italia: Dalla legge Casati del 1859 ai decreti delegati* (Milan: G. Mazzotta, 1975); Giampaolo Perugi, *Educazione e politica in Italia, 1860–1900* (Turin: Loescher, 1978); Ester De Port, *La scuola elementare dall'unità alla caduta del fascismo* (Bologna: Il Mulino, 1996).

68 Luigi Masulli Miglorini, *La sinistra storica al potere: Sviluppo della democrazia e direzione dello stato, 1876–1878* (Naples: La Guida, 1979); Giampiero Carocci, *Agostino Depretis e la politica interna italiana dal 1876 al 1887* (Turin: Einaudi, 1961).

69 *Atti della Giunta parlamentare per l'inchiesta agraria*, vol. 1, p. 111.

70 Ibid., vol. 8, p. 622.

71 Ibid., vol. 7, p. 236.

72 Ibid., vol. 13, p. 520.

73 Ibid., vol. 7, p. 191.

74 Ibid., vol. 1, p. 111.

75 Ibid., vol. 15, pp. 84–5.

3 WORKERS OF THE WORLD, 1870–1914

1 Antonio Lazzarini, *Campagne venete*, pp. 51f., quoted in René Del Fabbro, *Transalpini: Italienische Arbeitswanderung nach Süddeutschland im Kaiserreich 1870–1918* (Osnabrück: Universitätsverlag Rasch, 1996), p. 56.

2 Comparisons of Italy's emigration statistics with those of receiving countries suggest that Italy's figures (cited here) may overestimate actual border crossings between 1902 and 1910, for they reflect requests for passports, not actual departures. See George Calfut, "An Analysis of Italian Emigration Statistics, 1876–1914," *Jahrbuch für Geschichte von Staat, Wirtschaft und Gesellschaft Lateinamerikas* 14 (1977): 310–31; Riccardo Faini and Alessandra Venturini, "Italian Emigration in the Prewar Period," in Timothy J. Hatton and Jeffrey G. Williamson (eds), *Migration and the International Labor Market, 1850–1939* (London: Routledge, 1994), pp. 72–90. A fuller treatment of the problem of the statistics for the years around 1900 is Timothy J. Hatton and Jeffrey G. Williamson, *The Age of Mass Migration: Causes and Economic Impact* (New York: Oxford University Press, 1998), pp. 97–8.

3 In addition to Robert Foerster, *The Italian Emigration of Our Times* (New York: Russell & Russell, 1968), and Gianfausto Rosoli (ed.), *Un secolo di emigrazione italiana, 1876–1976* (Rome: Centro Studi Emigrazione, 1978), standard overviews of the mass migrations are in Franca Assante (ed.), *Movimento migratorio italiano dall'unità nazionale ai giorni nostri*, 2 vols. (Geneva: Librairie Droz, 1978); Zeffiro Ciuffoletti, *L'emigrazione nella storia d'Italia, 1868–1975; Storia e documenti* (Florence: Vallecchi, 1978); Ercole Sori, *L'emigrazione italiana dall'unità alla seconda guerra mondiale* (Bologna: Il Mulino, 1979); Renzo de Felice (ed.), *Cenni storici sull'emigrazione italiana* (Milan: Franco Angeli, 1979).

4 Immanuel Wallerstein, *The Modern World System II: Mercantilism and the Consolidation of the European World-Economy, 1600–1750* (New York: Academic Press, 1980); see also Alan K. Smith, *Creating a World Economy: Merchant Capital, Colonialism, and World Trade, 1400–1825* (Boulder, Colo.: Westview Press, 1991).

5 The concept of an imperial Atlantic economy has been widely used by European and American students of the slave trade and of the era before American nation-state formation. See, e.g., Philip D. Curtin, *The Rise and Fall of the Plantation Complex: Essays in Atlantic History* (Cambridge: Cambridge University Press, 1990).

6 The concept of an Atlantic economy has also been widely used by historians of migration of the nineteenth century, but without the same resonance as in the historiography of the colonial era. See Dirk Hoerder, "An Introduction to Labor Migration in the Atlantic Economies," in *Labor Migration in the Atlantic Economies; The European and North American Working Classes During the Period of Industrialization* (Westport, Conn.: Greenwood Press, 1985), pp. 6–9. Hoerder's work builds on the insights of Brinley Thomas, *Migration and Economic Growth; A Study of Great Britain and the Atlantic Economy*, 2nd edn (Cambridge: Cambridge University Press, 1973), and Michael Piore, *Birds of Passage: Migrant Labor in Industrial Societies* (Cambridge: Cambridge University Press, 1979). See also Bruno Ramirez, *On the Move: French-Canadian and Italian Migrants in the*

North Atlantic Economy, 1860–1914 (Toronto: McClelland and Stewart, 1991). A concept of an Atlantic economy also shapes the comparative work of Walter Nugent, *Crossings: The Great Transatlantic Migrations, 1870–1914* (Bloomington: Indiana University Press, 1992), and (from a very different methodological direction) Paul Gilroy, *The Black Atlantic: Modernity and Double Consciousness* (London: Verso, 1993).

7 Quoted in Jose C. Moya, *Cousins and Strangers: Spanish Immigrants in Buenos Aires, 1850–1930* (Berkeley: University of California Press, 1998), p. 49. See also Walter Nugent, "Frontiers and Empires in the Late Nineteenth Century," *Western Historical Quarterly* 20 (November 1989): 393–408.

8 Long fascinated with the nationalism and imperialism of the era, historians have only begun to interpret this earlier moment of economic globalization. Students of Pacific migrations and commerce now offer their perspectives on the globalization of this period. See Lucie Cheng and Edna Bonacich (eds), *Labor Immigration under Capitalism: Asian Workers in the United States Before World War II* (Berkeley: University of California Press, 1984). On globalization within Europe, see Carl Strikwerda, "The Troubled Origins of European Economic Integration: International Iron and Steel and Labor Migration in the Era of World War I," *American Historical Review* 98 (1993): 1106–42.

9 For worldwide migrations, see Lydia Potts, *The World Labour Market: A History of Migration* (London: Zed Books, 1990); Robin Cohen (ed.), *The Cambridge Survey of World Migration* (Cambridge: Cambridge University Press, 1995); Robin Cohen, *Global Diasporas, An Introduction* (Seattle: University of Washington Press, 1997). See also Dirk Hoerder, *Global Migrations* (Duke University Press, forthcoming 2001).

10 Leslie Page Moch, *Moving Europeans: Migration in Western Europe since 1650* (Bloomington: Indiana University Press, 1992), ch. 4.

11 Besides Potts, *The World Labour Market*; see also Stephen Castles and Mark J. Miller, *The Age of Migration: International Population Movements in the Modern World* (New York: The Guilford Press, 1993), ch. 3. Migrations of Europeans have received a disproportionate share of scholarly attention. See the seminal essay: Frank Thistlethwaite, "Migration from Europe Overseas in the Nineteenth and Twentieth Centuries," in Rudolph J. Vecoli and Suzanne M. Sinke (eds), *A Century of European Migrations, 1830–1930* (Urbana: University of Illinois Press, 1991), pp. 17–49 (1st edn 1960); J.D. Gould, "European Inter-Continental Emigration, 1815–1914: Patterns and Causes," *Journal of European Economic History* 8 (Winter 1979): 593–679.

12 George E. Pozzetta (ed.), *Pane e Lavoro: The Italian American Working Class* (Toronto: Multicultural History Society of Ontario, 1980).

13 See Introduction, p. 3, note 6.

14 Ireland's population, unlike Italy's, actually declined: see Nugent, *Crossings*, Tables 1 and 4.

15 There is no comprehensive analysis of Chinese global migrations. Helpful are: Linda Pan, *Sons of the Yellow Emperor: The Story of the Overseas Chinese* (London: Mandarin, 1991); Gungwu Wang, *China and the Chinese Overseas* (Singapore: Times Academic Press 1991); Dudley L. Poston, Jr. and Yu Mei-Yu, "International Migration: The Distribution of the Overseas

Chinese," in Dudley L. Poston, Jr. and David Yaukey, *The Population of Modern China* (New York: Plenum Press, 1992).

16 Rudolph J. Vecoli, "The Italian Diaspora, 1876–1976," in Robin Cohen (ed.), *The Cambridge Survey of World Migration* (Cambridge: Cambridge University Press, 1995), p. 114; Foerster, *The Italian Emigration of Our Times*, pp. 3–4; see also Mark Wyman, *Round-Trip to America: The Immigrants Return to Europe, 1880–1930* (Ithaca, N.Y.: Cornell University Press, 1993), and J.D. Gould, "Emigration: The Road Home: Return Migration from the U.S.A.," *Journal of European Economic History* 9 (Spring 1980): 41–112.

17 *Annuario statistico della emigrazione italiana* (Rome: Commissariato dell'Emigrazione, 1927), pp. 241–2; Samuel L. Baily, *Immigrants in the Lands of Promise: Italians in Buenos Aires and New York City, 1870–1914* (Ithaca, N.Y.: Cornell University Press, 1999), Table 2.3.

18 My analysis of 408 elite migrants in Ugo E. Imperatori, *Dizionario di italiani all'estero* (Genoa: L'Emigrante, 1956).

19 Patrizia Audenino, *Un mestiere per partire* (Milan: Franco Angeli, 1990).

20 *Atti della Giunta parlamentare per l'inchiesta agraria* (Rome: Forzani, 1881–6), vol. 7, fasc. 1, p. 191. See also Dino Cinel, "The Seasonal Emigration of Italians in the Nineteenth Century: From Internal to International Destinations," *Journal of Ethnic Studies* 10, 1 (1982): 43–68.

21 Michael La Sorte, *La Merica; Images of Italian Greenhorn Experience* (Philadelphia, Pa.: Temple University Press, 1985), p. 6.

22 John Potestio, "Le memorie di Giovanni Veltri: Da contadino a impresario di ferrovie," *Studi emigrazione* 22 (March 1985): 129–40.

23 From an expanding literature, see Robin Cohen, *The New Helots, Migrants in the International Division of Labour* (Brookfield, Vt.: Gower, 1988), ch. 1; Tom Brass and Marcel van der Linden, *Free and Unfree Labour* (Amsterdam: International Institute for Social History, 1993); Robert J. Steinfeld, *The Invention of Free Labor: The Employment Relation in English and American Law and Culture, 1350–1870* (Chapel Hill: The University of North Carolina Press, 1991).

24 Karl Marx, *Capital: A Critique of Political Economy* (Harmondsworth: Penguin Books, 1976), vol. 1, p. 925.

25 Amy A. Bernardy, *Italia randaggia; Attraverso gli Stati Uniti* (Turin: Fratelli Bocca, 1913), p. 84.

26 See the contract in Mario Enrico Ferrari, "I mercanti di fanciulli nelle campagne e la tratta dei minori: Una realtà sociale dell'Italia fra '800 e '900," *Movimento operaio e socialista* 6 (1983): 107–8; John Zucchi, "'Les petits Italiens': Italian Child Street Musicians in Paris, 1815–1875," *Studi emigrazione* 97 (1990): 27–54.

27 John E. Zucchi, *The Little Slaves of the Harp: Italian Child Street Musicians in Nineteenth-Century Paris, London, and New York* (Montreal: McGill-Queen's University Press, 1992), p. 133.

28 But see Zeffiro Ciuffoletti, "Sfruttamento della manodopera infantile italiana in Francia alla fine del sec. XIX," *Affari sociali internazionali* 5 (Fall–Winter 1977): 250.

29 Robert F. Harney, "Montreal's King of Italian Labour: A Case Study of Padronism," *Labour/Le Travail* 4 (1979): 57–84.

30 Humbert S. Nelli, "The Italian Padrone System in the United States," *Labor History* 2 (Spring 1964): 153–67; Robert F. Harney, "The Padrone

and the Immigrant," *Canadian Review of American Studies* 5 (Fall 1974): 101–18; Robert Harney, "The Commerce of Migration," *Canadian Ethnic Studies* 9, 1 (1977): 42–53; Luciano J. Iorizzo, *Italian Immigration and the Impact of the Padrone System* (New York: Arno Press, 1980). The significance of the charges against padroni and their slaves long shaped the U.S. historiography. See Rudolph J. Vecoli, "Italian American Workers, 1880-1920: Padrone Slaves or Primitive Rebels?," in Silvano Tomasi (ed.), *Perspectives in Italian Immigration and Ethnicity* (New York: Center for Migration Studies, 1975), pp. 25–50; Donna Gabaccia, "Neither Padrone Slaves nor Primitive Rebels: Sicilians on Two Continents," in Dirk Hoerder (ed.), *"Struggle a Hard Battle": Essays on Working-Class Immigrants* (DeKalb: Northern Illinois University Press, 1986), pp. 95–120.

31 Elda Zappi, *If Eight Hours Seem Too Few: Mobilization of Women Workers in the Italian Rice Fields* (Albany: State University of New York Press, 1991), pp. 42–3.

32 Foerster, *The Italian Emigration of Our Times*, pp. 131–2.

33 Peter Way, *Common Labour: Workers and the Digging of North American Canals, 1780–1860* (Cambridge: Cambridge University Press, 1993); see also Terry Coleman, *The Railway Navvies: A History of the Men Who Made the Railways* (London: Hutchinson, 1965), ch. 3.

34 *The World in My Hand, Italian Emigration in the World 1860/1960* (Rome: Centro Studi Emigrazione, 1997), pp. 63–80, passim.

35 La Sorte, *La Merica*, p. 76.

36 Persia Crawford Campbell, *Chinese Coolie Emigration to Countries within the British Empire* (London: P.S. King & Son, 1923).

37 From an extensive literature, Anthony H. Galt, "Re-thinking Patron–Client Relationships: The Real System and the Official System in Southern Italy," *Anthropological Quarterly* 47 (1974): 182–202; Jeremy Boissevain, "Patronage in Sicily," *Man* 1 (1966): 18–33. Many patrons in the South were what Jane and Peter Schneider have called "broker capitalists:" see their *Culture and Political Economy in Western Sicily* (New York: Academic Press, 1976), pp. 170–2.

38 Fernando Manzotti, *La polemica sull'emigrazione nell'Italia unita* (Milan: Dante Alighieri, 1969), p. 16.

39 See Chapter 2, p. 56, above.

40 M.F. Zuleika Alvim, *Brava gente! Os italianos em São Paulo, 1870–1920* (São Paulo: Brasiliense, 1986), pp. 54–100; Angelo Trento, *La dovè la raccolta del caffè: L'emigrazione italiana in Brasile 1875–1940* (Padua: Antenore, 1984); Chiara Vangelista, *Le braccia per la fazenda* (Milan: Franco Angeli, 1982). English-only readers may consult: Thomas H. Holloway, *Immigrants on the Land: Coffee and Society in São Paulo, 1886–1934* (Chapel Hill: University of North Carolina Press, 1980); and Thomas H. Holloway, "Creating the Reserve Army: The Immigration Program of São Paulo, 1886–1930," *International Migration Review* 12 (Summer 1978): 187–209; see also Warren Dean, *Rio Claro: A Brazilian Plantation System, 1820–1920* (Stanford, Calif.: Stanford University Press, 1976). On the Italian state's suspension of assisted migration, see Chapter 6, p. 136, below. As with the child apprentices, the government ignored peonage in Italy.

41 William A. Douglass, *From Italy to Ingham: Italians in North Queensland* (St. Lucia: University of Queensland Press, 1995), ch. 3.
42 J.S. MacDonald, "Chain Migration, Ethnic Neighborhood Formation and Social Networks," *Millbank Memorial Fund Quarterly* 42 (January 1964): 82–91; see also Franc Sturino, *Forging the Chain: A Case Study of Italian Migration to North America, 1880–1930* (Toronto: Multicultural History Society of Ontario, 1990); Samuel L. Baily, "Chain Migration of Italians to Argentina: Case Studies of the Agnonesi and the Sirolesi," *Studi emigrazione* 19 (1982): 73–91; Fernando J. Devoto, "Álgo mas sobre las cadenas migratoris de los italianos a la Argentina," *Estudios migratorios Latinoamericanos* 19 (1991): 335–6.
43 U.S. Senate, Immigration Commission, *Reports* (Washington, D.C.: Government Printing Office, 1911), vol. 4, p. 59.
44 *Emigrazione e colonizzazione nelle lettere dei contadini veneti e friulani in America Latina 1876–1902* (Verona: Cierre, 1994), p. 187.
45 Donna Gabaccia, *Militants and Migrants: Rural Sicilians Become Italian Workers* (New Brunswick: Rutgers University Press, 1988); Samuel L. Baily and Franco Ramella (eds), *One Family, Two Worlds: An Italian Family's Correspondence across the Atlantic, 1901–1922* (New Brunswick: Rutgers University Press, 1988), pp. 10–12.
46 An early effort to explore the relation between a village and its diaspora is Joseph Lopreato and D. Lococo, "Stefanaconi: Un villaggio agricolo meridionale in relazione al suo 'mondo'," *Quaderni di sociologia* 34 (Autumn 1959): 239–60.
47 *Annuario statistico*, Table V.
48 J.S. MacDonald, "Some Socio-Economic Emigration Differentials in Rural Italy, 1902–1913," *Economic Development and Cultural Change* 7 (1958): 55–72; J.S. MacDonald, "Agricultural Organization, Migration and Labour Militancy in Rural Italy," *Economic History Review* 2nd ser. 16 (1963): 61–75.
49 M. Triaca, *Amelia: A Long Journey* (Richmond: Victoria, 1985), p. 35, quoted in Richard Bosworth, *Italy and the Wider World* (New York: Routledge, 1996), p. 134.
50 Dirk Hoerder, "From Dreams to Possibilities: The Secularization of Hope and the Quest for Independence," in Dirk Hoerder and Horst Rössler (eds), *Distant Magnets: Expectations and Realities in the Immigrant Experience, 1840–1930* (New York: Holmes & Meier, 1993), puts the lure of the U.S. in comparative perspective.
51 *Atti della Giunta parlamentare per l'inchiesta agraria*, vol. 3, fasc. 2, p. 620; vol. 7, fasc. 2, p. 346; vol. 8, tome 1, fasc. 2, p. 74.
52 Ibid., vol. 6, fasc. 1, pp. 44–5.
53 Gabaccia, *Militants and Migrants*, pp. 83–4.
54 Donna Gabaccia and Franca Iacovetta, "Women, Work, and Protest in the Italian Diaspora: Gendering Global Migration, Rethinking Family Economies, Nationalisms, and Labour Activism," *Labour/Le Travail* 42 (Fall 1998): 161–81.
55 Paola Corti, "I movimenti frontalieri al femminile; Percorsi tradizionali ed emigrazione di mestiere dalle valli cuneesi alla Francia meridionale," special issue "L'esodo frontaliero: Gli italiani nella Francia meridionale," *Recherches régionales: Alpes-Maritimes et contrées limitrophes* (1995): 65–90.

56 Del Fabbro, *Transalpini*, p. 4.
57 Rosoli, *Un secolo di emigrazione italiana*, Table XX.
58 Baily, *Immigrants in the Lands of Promise*, Table 3.4.
59 Betty Boyd Caroli, *Italian Repatriation from the United States, 1900–1914* (New York: Center for Migration Studies, 1973), pp. 49–50.
60 Emilio Franzina, *Merica! Merica!: Emigrazione e colonizzazione nelle lettere dei contadini veneti e friulani in America Latina 1876–1902* (Verona: Cierre, 1994), pp. 190–1.
61 For the early autonomy movement in Sicily, see Michele Vaina, *Popolarismo e nasismo in Sicilia* (Florence: La Rinascità del Libro, 1911); an historian's assessment is Enrico La Loggia, *Autonomia e rinascità della Sicilia* (Palermo: Ires, 1953).
62 Giorgio Colombo, *La scienza infelice: Il museo di antropologia criminale di Cesare Lombroso* (Turin: B. Boringhieri, 1975). See also D. Pick, "The Faces of Anarchy: Lombroso and the Politics of Criminal Science in Post-Unification Italy," *History Workshop* 21 (1986): 60–86.
63 The image comes from Franca Iacovetta, *Such Hardworking People: Italian Immigrants in Postwar Toronto* (Montreal: McGill-Queen's University Press, 1992).
64 There are few general studies. Italian contractors did undertake work in China, as the *Bollettino* suggested; see Audenino, *Un mestiere per partire*, illustrations, pp. 17–18. For background, Nicholas Faith, *The World the Railways Made* (London: The Bodley Head, 1990), and Leone Carpi, *Delle colonie e dell'emigrazione d'italiani all'estero sotto l'aspetto dell'industria, commercio ed agricoltura* (Milan: Lombarda, già D. Salvi, 1874), vols. 1–2. On the construction industry and migratory workers from the North, see Elisabetta Calderini, Rocco Curto, and Gemma Sirchia, *Hirondelles, 1860–1914: Storia e vicende dei lavoratori dell'edilizia in Piemonte* (Turin: Celid, 1985); Patrizia Audenino, "The Paths of the Trade: Italian Stone Masons in the United States," *International Migration Review* 20 (1986): 779–95. For railway workers in Canada, see Bruno Ramirez, "Brief Encounters: Italian Immigrant Workers and the CPR 1900–1930," *Labour/Le Travail* 17 (Spring 1986): 9–27. The best overview remains Foerster, *The Italian Emigration of Our Times*, chs. 8–11, 14, 18.
65 Compare Douglass, *From Italy to Ingham*, ch. 3; Jean Scarpaci, *Italian Immigrants in Louisiana's Sugar Parishes: Recruitment, Labor Conditions, and Community Relations, 1880–1910* (New York: Arno Press, 1980); Jean Scarpaci, "Labor for Louisiana's Sugar Fields: An Experiment in Immigrant Recruitment," *Italian Americana* 7 (1981): 19–41; Jean Scarpaci, "Immigrants in the New South: Italians in Louisiana's Sugar Parishes, 1880–1910," *Labor History* 16 (1975): 165–83.
66 James R. Scobie, *Revolution on the Pampas: A Social History of Argentine Wheat, 1860–1910* (Austin: University of Texas Press, 1964); Carina Frid de Silberstein, "Labor and Migration in an Agricultural Economy: Italians in Argentina," in Donna Gabaccia and Fraser Ottanelli (eds), *For Us There Are no Frontiers* (forthcoming).
67 Foerster, *The Italian Emigration of Our Times*, pp. 131–4.
68 Except as padroni, the migratory entrepreneurs have not yet found their historian. See John Zucchi, "Occupations, Enterprise and the Migration Chain: The Fruit Traders from Termini Immerese in Toronto, 1900–

1930," *Studi emigrazione* 77 (March 1985); Donna Gabaccia, "Italians in American Food Industries," *Italian American Review* 6, 2 (Autumn 1997): 1–19; Sebastian Fichera, "Entrepreneurial Behavior in an Immigrant Colony; The Economic Experience of San Francisco's Italian-Americans, 1850–1940," *Studi emigrazione* 32, 118 (1995): 321–45.

69 Besides Foerster, *The Italian Emigration of Our Times*, see Serge Étienne Kagan Bonnet and Michel Maigret, *L'Homme du fer; Mineurs de fer et ouvriers sidérurgistes lorrains, 1890–1930*, 2 vols. (Nancy: Centre Lorrain d'Études Sociologiques, n.d. but circa 1974 or 1975); Del Fabbro, *Transalpini*, pp. 161–7; Robert Pascoe and Patrick Bertola, "Italian Miners and the Second-Generation 'Britishers' at Kalgoorlie, Australia," *Social History* 10 (January 1985): 936.

70 Nancy G. Eschelmann, "Forging a Socialist Women's Movement: Angelica Balabanoff in Switzerland," in Betty Boyd Caroli, Robert F. Harney and Lydio F. Tomasi (eds), *The Italian Immigrant Woman in North America* (Toronto: Multicultural History Society of Ontario, 1978), pp. 44–75; Silvia Corazza, "Itinerari professionali femminili: Le setaiole di una comunità manifatturiera piemontese nella Francia meridionale," in special issue, "L'esodo frontaliero," *Recherches régionales*, pp. 107–17.

71 From a sizeable literature on the U.S., Miriam Cohen, *Workshop to Office: Two Generations of Italian Women in New York City, 1900–1950* (Ithaca, N.Y.: Cornell University Press, 1993). Less has been written on Argentina, Brazil, and Canada but see the articles in the special issue "Le emigrate italiane in prospettiva comparata," *Altreitalie* 9 (1993): Franca Iacovetta, "Scrivere le donne nella storia dell'immigrazione: Il caso italo-canadese"; Loraine Slomp Giron, "L'immigrata in Brasile e il lavoro"; and Alicia Bernasconi and Carina Frid de Silberstein, "Le altre protagoniste: italiane a Santa Fe." See also Carina Frid de Silberstein, "Becoming Visible: Italian Immigrant Women in the Garment and Textile Industries in Argentina, 1890–1930," in Donna Gabaccia and Franca Iacovetta, *Foreign, Female and Fighting Back*, forthcoming. On Brazil, see also Zuleika Alvim, "Lavoro femminile ed economia domestica nelle fazendas italiane di S. Paulo all'inizio del secolo," *Studi emigrazione* 20 (1983): 237–46; Zuleika Alvin, "Immigrazione e forza lavoro femminile a São Paulo, 1830–1920," in Emilio Franzina (ed.), *Un altro Veneto, Saggi e studi di storia dell'emigrazione dei secoli XIX e XX* (Abano Terme: Francisci, 1983), pp. 492–512.

72 On male industrial employment worldwide, see the essays collected in Bruno Bezza, *Gli italiani fuori d'Italia, Gli emigrati italiani nei movimenti operai dei paesi d'adozione 1880–1940* (Milan: Franco Angeli, 1983).

73 Baily, *Immigrants in the Land of Promise*, pp. 38–9. For an introduction to the rather large literature on domestic service, Gabaccia and Iacovetta, "Women, Work, and Protest in the Italian Diaspora."

74 Hugh Tinker, *A New System of Slavery: The Export of Indian Labour Overseas, 1830–1920* (London: Oxford University Press, 1974).

75 W. Kloosterboer, *Labour Since the Abolition of Slavery; A Survey of Compulsory Labour Throughout the World* (Westport, Conn.: Greenwood Press, 1974), 1st edn 1960, remains the best overview. See also Jan Breman, *Labour Migration and Rural Transformation in Colonial Asia* (Amsterdam: Free University Press, 1990); Stanley L. Engerman, "Servants to Slaves to Servants:

Contract Labour and European Expansion," in Pieter C. Emmer and Magnus Mörner (eds), *European Expansion and Migration; Essays on the International Migration from Africa, Asia and Europe* (New York: Oxford University Press, 1992), pp. 263–94.

76 Donna Gabaccia, "The 'Yellow Peril' and the 'Chinese of Europe': Global Perspectives on Race and Labor, 1815–1930," in Jan and Leo Lucassen (eds), *Migrations, Migration History: Old Paradigms and New Perspectives* (New York: Berg and International Institute for Social History, 1997).

77 John W. Briggs, *An Italian Passage, Immigrants to Three American Cities, 1890–1930* (New Haven, Conn.: Yale University Press, 1978), pp. 112–15.

78 On Italians in northeastern cities of the U.S., Josef Barton, *Peasants and Strangers: Italians, Rumanians, and Slovaks in an American City, 1890–1950* (Cambridge, Mass.: Harvard University Press, 1975); Thomas Kessner, *The Golden Door: Italian and Jewish Immigrant Mobility in New York City, 1880–1915* (New York: Oxford University Press, 1977); Virginia Yans-McLaughlin, *Family and Community: Italian Immigrants in Buffalo, 1880–1930* (Ithaca, N.Y.: Cornell University Press, 1977); John E. Bodnar, Roger D. Simon, and Michael P. Weber, *Lives of their Own: Blacks, Italians, and Poles in Pittsburgh, 1900–1960* (Urbana: University of Illinois Press, 1983); Humbert S. Nelli, *Italians in Chicago, 1880–1930; A Study in Ethnic Mobility* (New York: Oxford University Press, 1970); Anna Maria Martellone, *Una little Italy nell'Atene d'America, La comunità italiana di Boston dal 1800 al 1920* (Naples: Guida, 1973); William DeMarco, *Ethnics and Enclaves: Boston's Italian North End* (Ann Arbor: UMI Press, 1981).

79 Philip F. Notarianni, "Italian Involvement in the 1903-04 Coal Miners' Strike in Southern Colorado and Utah," and Betty Boyd Caroli, "Italians in the Cherry, Illinois, Mine Disaster," in Pozzetta, *Pane e Lavoro*, pp. 47–65 and 67–79; Dolores Manfredini, "The Italians Come to Herrin," *Journal of the Illinois State Historical Society* (December 1944): 317–19. See also the works of Bonifacio Bolognani: *Il pane della miniera: Speranze, sacrifici e morte di emigrati trentini in terra d'America* (Trento: Bernardo Clesio, 1988), and *Del Trentino alla terra del diamante nero* (Trento: Bernardo Clesio, 1986).

80 But most work has focused on the urban Italians of San Francisco; see Rose Scherini, *The Italian American Community of San Francisco: A Descriptive Study* (New York: Arno Press, 1980); Dino Cinel, *From Italy to San Francisco: The Immigrant Experience* (Stanford, Calif.: Stanford University Press, 1982); Deanna Paoli Gumina, *The Italians of San Francisco, 1850–1930* (New York: Center for Migration Studies, 1985).

81 Briggs, *An Italian Passage*, ch. 8.

82 Gary R. Mormino and George E. Pozzetta, *The Immigrant World of Ybor City: Italians and Their Latin Neighbors in Tampa, 1885–1985* (Urbana: University of Illinois Press, 1987); see also Ferdinando Fasce, *Una famiglia a stelle e strisce; Grande guerra e cultura d'impresa in America* (Bologna: Il Mulino, 1993).

83 Besides Fasce, *Una famiglia*, see Briggs, *Italian Passage*, Table 5.29; Kessner, *The Golden Door*, Tables 3 and 8.

84 The best introductions to the work of Italians in Canada are in John Zucchi, *Italians in Toronto: Development of a National Identity 1875–1935* (Montreal and Kingston: McGill-Queen's University Press, 1988), ch. 8;

Bruno Ramirez and Michael Del Balso, *The Italians of Montreal: From Sojourning to Settlement 1900–1921* (Montreal: Associazione di Cultura Popolare Italo-Quebecchese, 1980); Roberto Perin and Franc Sturino (eds), *Arrangiarsi: The Italian Immigration Experience in Canada* (Montreal: Guernica, 1992).

85 For an introduction, see Gianfranco Cresciani, *The Italians* (Sydney: Australian Broadcasting Corporation, 1985) and Gianfranco Cresciani, *Australia, the Australians and the Italian Migration* (Milan: Franco Angeli, 1983). Also useful are: the older work of W.D. Borrie, *Italians and Germans in Australia* (Melbourne: The Australian National University, 1954); and the more popular account, Tito Cecilia, *We Didn't Arrive Yesterday: Outline of the History of the Italian Migration into Australia from Discovery to the Second World War* (Red Cliffs: Scalabrinians, 1987).

86 Kay Saunders (ed.), *Indentured Labour in the British Empire 1834–1920* (London: Croom Helm, 1984). Unlike Italians, Chinese found a small place in the British Empire; see Look Lai Walton, *Indentured Labor, Caribbean Sugar: Chinese and Indian Migrants to the British West Indies* (Baltimore, Md.: The Johns Hopkins University Press, 1993); Peter Richardson, *Chinese Mine Labor in the Transvaal* (London: Macmillan Press, 1982).

87 Lynn Lees, *Exiles of Erin: Irish Migrants in Victorian London* (Ithaca, N.Y.: Cornell University Press, 1979).

88 Lucio Sponza, *Italian Immigrants in Nineteenth-Century Britain: Realities and Images* (Leicester: Leicester University Press, 1988); see also "I gelatieri in Gran Bretagna," *Italiani nel mondo*, 31 (1975).

89 Núncia Santoro de Constantino, *O italiano da esquina; Imigrantes na sociedade porto-alegrense* (Porto Alegre: Escola Superior de Teologia e Espiritualidade Franciscana, 1991). On Argentina, see Ezequiel Gallo, "Los Italianos en los Origines de la Agricultura Argentina: Santa Fe (1870–1895)," in Francis Korn (ed.), *Los Italianos en la Argentina* (Buenos Aires: Fondazione Giovanni Agnelli, 1983); Eugenia Scarzanella, "Immigrazione italiana e colonizzazione agricola in Argentina (1860–1880)," in Renzo de Felice, *Cenni storici sulla emigrazione italiana nelle Americhe e in Australia* (Milan: Franco Angeli, 1979); Alicia Bernasconi, "Inmigración italiana, colonización y mutualismo en el Centro-norte de la provincia de Santa Fe," in Fernando J. Devoto and Gianfausto Rosoli (eds), *L'Italia nella società argentina* (Rome: Centro Studi Emigrazione, 1988), pp. 178–89.

90 James R. Scobie, *Buenos Aires, Plaza to Suburb, 1870–1910* (New York: Oxford University Press, 1974); Gino Germani, "Mass Immigration and Modernization in Argentina," in Irving L. Horowitz (ed.), *Masses in Latin America* (New York: Oxford University Press, 1970), pp. 289–330.

91 On middle-class migrants, see Carina Frid de Silberstein, "Parenti, negozianti e dirigenti: La prima dirigenza italiana di Rosario (1860–1890)," in Gianfausto Rosoli (ed.), *Identità degli italiani in Argentina* (Rome: Studium, 1993), pp. 129–65; Romolo Gandolfo, "Notas sobre la elite de una comunidad emigrada en cadena: el caso de los Agnoneses," in Devoto and Rosoli, *L'Italia nella società argentina*, pp. 160–77; María Inés Barbero and Susana Felder, "El rol de los italianos en el nacimiento y desarrollo de las asociaciones empresarias en la Argentina (1880–1930)," in Devoto and Rosoli, *L'Italia nella società argentina*, pp. 137–59.

92 Baily, *Immigrants in the Lands of Promise*, pp. 73–5; see also Herbert S. Klein, "The Integration of Italian Immigrants into the United States and Argentina: A Comparative Analysis," *American Historical Review* 88 (1983): 281–305.

93 From a large literature, see Baily, *Immigrants in the Lands of Promise*, ch. 5; Mirta Z. Lobato, "Una visión del mundo del trabajo; Obreros inmigrantes en la industria frigorífica 1900–1930," in Fernando J. Devoto and Eduardo J. Miguez (eds), *Asociacionismo, trabajo e identidad etnica; Los italianos en América Latina en una perspectiva comparada* (Buenos Aires: CEMLA-CSER-IEHS, 1992), pp. 189–204; Edgar Rodrigues, *Lavoratori italiani in Brasile: Un secolo di storia sociale dell'altra Italia* (Casavelino Scala: Galzerano Editore, 1985); Edgar Rodrigues, *Os anarquistas: Trabalhadores italianos no Brasil* (São Paulo: Global, 1984); Michael Hall, "Immigration and the Early São Paulo Working Class," *Jahrbuch für Staat, Wirtschaft und Gesellschaft Lateinamerikas* 2 (1975): 393–407.

94 For a review of the sizeable literature, see Alejandro Portes, "The Ethnic Enclave Debate Revisited," *International Journal of Urban and Regional Research* 17, 3 (1993): 428–36.

95 Torcuato Di Tella, "Argentina: Un'Australia italiana? L'impatto dell'emigrazione sul sistema politico argentina," in Bezza, *Gli italiani fuori d'Italia*, pp. 419–54.

96 Hans-Martin Habicht, *Probleme der italienischen Fremdarbeiter im Kanton St. Gallen vor dem Ersten Weltkrieg* (Herisau: S&S, 1977); Madelyn Holmes, *Forgotten Migrants, Foreign Workers in Switzerland before World War I* (Rutherford, N.J.: Fairleigh Dickinson University Press, 1988).

97 The best general introductions are Michel Dreyfus and Pierre Milza, *Un siècle d'immigration italienne en France, 1850–1950* (Paris: CEDEI, 1987); Jean Baptiste Duroselle and Enrico Serra (eds), *L'emigrazione italiana in Francia prima del 1914* (Milan: Franco Angeli, 1978); Pierre Milza, "L'émigration italienne en France de 1870 à 1914," *Affari sociali internazionali* 5 (1977): 63–86.

98 Besides Del Fabbro, *Transalpini*, see René Del Fabbro, "Italienische Industriearbeiter im Wilhelminischen Deutschland (1890–1914)," *Vierteljahrschrift für Sozial- und Wirtschaftsgeschichte* 76 (1989): 202–28; Hermann Schäfer, "Italienische 'Gastarbeiter' im deutschen Kaiserreich (1890–1914)," *Zeitschrift für Unternehmensgeschichte* 27 (1982): 192–214; Ulrich Herbert, *A History of Foreign Labor in Germany* (Ann Arbor: University of Michigan Press, 1984); Adolf Wennemann, *Die Italiener im Rheinland und Westfalen des späten 19. und frühen 20. Jahrhunderts* (Osnabrück: IMS Schriften, 1996). On Austria, see Marina Cattaruzza, *La formazione del proletariato urbano; Immigrati, operai di mestiere, donne a Trieste dalla metà del secolo XIX alla prima guerra mondiale* (Turin: Musolini Ed., 1978).

4 TRANSNATIONALISM AS A WAY OF WORKING-CLASS LIFE

1 Giuseppe Pitrè, *Proverbi siciliani* (Palermo: "Il Vespro," 1978), first published 1870–1913, vol. 3, p. 118.

2 Antonio Gibelli, "La risorsa America," in Antonio Gibelli and Paride Rugafiori (eds), *La Liguria* (Turin: G. Einaudi, 1994), pp. 585–650.

3 A large scholarly literature has long debated the supposed familism of southern Italians. See, e.g., the work of Edward Banfield, *The Moral Basis of a Backward Society* (Glencoe, Ill.: The Free Press, 1958). His many critics include: Sydel F. Silverman, "Agricultural Organization, Social Structure and Values in Italy: Amoral Familism Reconsidered," *American Anthropologist* 70 (1968): 1–20; William Muraskin, "The Moral Basis of a Backward Sociologist: Edward Banfield, the Italians and the Italian-Americans," *American Journal of Sociology* 79 (1974): 1484–96. For a general historical introduction to the issues of Italian family history, see Paolo Macry, "Rethinking a Stereotype: Territorial Differences and Family Models in the Modernization of Italy," *Journal of Modern Italian Studies* 2, 2 (1997): 188–214; David I. Kertzer and Richard P. Saller (eds), *The Family in Italy: From Antiquity to the Present* (New Haven, Conn.: Yale University Press, 1991).

4 On peasants, Karl Marx, *The Eighteenth Brumaire of Louis Bonaparte* (New York: International Publishers, 1957), p. 109, and David Mitrany, *Marx Against the Peasant* (New York: Collier, 1961). The best statement of Marx on internationalism is his "Critique of the Gotha Program," in *The Marx–Engels Reader*, ed. Robert C. Tucker (New York: W.W. Norton, 1972), pp. 390–1. Marx's phrase "workers of the world" is, of course, from the *Communist Manifesto*.

5 A major challenge, however, is the very preliminary state of scholarship on Italy's rural women. In English, see Emiliana Noether, "The Silent Half: Le Contadine del Sud before World War One," in Betty Boyd Caroli, Robert F. Harney, and Lydio F. Tomasi (eds), *Italian Immigrant Women in North America* (Toronto: Multicultural History Society of Ontario, 1978); Donna Gabaccia, "In the Shadows of the Periphery: Italian Women in the Nineteenth Century," in Marilyn J. Boxer and Jean H. Quataert (eds), *Connecting Spheres: Women in the Western World, 1500 to the Present* (New York: Oxford University Press, 1987). Good starting places for those interested in women and gender in Italian rural society are: Nuto Revelli, *L'anello forte; La donna; Storie di vita contadina* (Turin: Einaudi, 1985); Paola Corti, "Società rurale e ruoli femminili in Italia tra Otto e Novecento," *Annali dell'Istituto Alcide Cervi* (special issue) 12 (1994). See also Jole Calapso, "La donna in Sicilia e in Italia: La realtà e la falsa coscienza nella statistica dal 1871 ad oggi," *Quaderni siciliani* 2 (March–April 1973): 13–20.

6 There is a large, and contentious, scholarly literature on family economies. For the evolution of the argument, particularly about the relationship of patriarchy, gender, and family solidarity and decision-making, see: Louise A. Tilly and Joan Scott, *Women, Work and Family* (New York: Holt, Rinehart and Winston, 1978); Patricia Hilden, "Family History vs. Women in History: A Critique of Tilly and Scott," *International Labor and Workingclass History* 16 (Fall 1979): 1–11; Louise Tilly, "After Family Strategies, What?," *Historical Methods* 20 (Summer 1987): 123–5.

7 Robert F. Harney, "Men Without Women," in Caroli *et al.*, *The Italian Immigrant Woman.*

8 Luigi Villari, *Gli Stati Uniti d'America e l'emigrazione italiana* (Milan, 1912), pp. 204–5, quoted in Richard Bosworth, *Italy and the Wider World* (New York: Routledge, 1996), p. 120.

9 Bruno Ramirez, *On the Move: French-Canadian and Italian Migrants in the North Atlantic Economy, 1860–1914* (Toronto: McClelland & Stewart, 1991), p. 97.
10 See, e.g., *Inchiesta parlamentare sulle condizioni dei contadini nelle provincie meridionali e nella Sicilia* vol. 4: *Benevento* (Rome: Tip. Nazionale di Giovanni Bertero, 1909).
11 From an extensive literature see Antonio Marinoni, *Come ho "fatto" l'America* (Milan: Athena, 1932).
12 For rural Italian family economies, see Franco Ramella, *Terra e telai: Sistemi di parentela e manifattura nel Biellese dell'Ottocento* (Turin: Einaudi, 1984); Donna R. Gabaccia, *From Sicily to Elizabeth Street: Housing and Social Change Among Italian Immigrants, 1880–1930* (Albany: State University of New York Press, 1984); Fortunata Piselli, *Parentela ed emigrazione; Mutamenti e continuità in una comunità calabrese* (Turin: Einaudi, 1981); A. Berrino, "Famiglia, terra, ed emigrazione in una comunità della costiera sorrentina," Andreina De Clementi, "La prima emigrazione," and Gabriella Gribaudi, "Emigrazione e modelli familiari," all in Pasquale Villani and Paolo Macry (eds), *La Campania* (Turin: Einaudi, 1990), pp. 399–421, 373–96, 425–37.
13 *Atti della Giunta parlamentare per l'inchiesta agraria* (Rome: Forzani, 1881–6), vol. 3, fasc. 2, p. 656. In other parts of central Italy they were called reggitore and reggitrice, ibid., vol. 1, fasc. 1, pp. 237–9.
14 The key theoretical reading on the peasant stem family is Lutz Berkner, "The Stem-Family and the Developmental Cycle of the Peasant Household: An Eighteenth Century Austrian Example," *American Historical Review* 77 (1972): 398–418. See also David I. Kertzer, *Family Life in Central Italy 1880–1910: Sharecropping, Wage Labor, and Co-residence* (New Brunswick: Rutgers University Press, 1984). On joint families, see William A. Douglass, "A Joint-Family Household in Eighteenth Century Southern Italian Society," in Kertzer and Saller, *The Family in Italy*, pp. 286–303. In the 1870s and 1880s, the joint household of brothers — which a Jacini commissioner termed "a very curious type of patriarchal republic" — existed mainly in Val d'Aosta and other parts of Piedmont. See *Atti della Giunta parlamentare per l'inchiesta agraria*, vol. 8, fasc. 2, pp. 641, 656.
15 A good general source on changing peasant agriculture in the nineteenth century remains Emilio Sereni, *Il capitalismo nelle campagne, 1860–1900* (Turin: Einaudi, 1968). On the proletarianized agriculturalists of the Po Valley, see Secondo Giacobbi (ed.), *Braccianti e contadini nella Valle Padana, 1880–1905* (Rome: Riuniti, 1975). See also *Il mondo agrario tradizionale nella valle padana; Atti del convegno di studi sul folklore padano, Modena, 17–19 marzo 1962* (Florence: Olschki, 1963).
16 The flavor of South Italian everyday life can be glimpsed in English in Anton Blok, "South Italian Agrotowns," *Comparative Studies in Society and History* 11 (1969): 121–35; Charlotte Gower Chapman, *Milocca, a Sicilian Village* (Cambridge, Mass.: Schenkman, 1971); Gabaccia, *From Sicily to Elizabeth Street.*
17 Giuseppe Pitrè, *La famiglia, la casa, la vita del popolo siciliano*, vol. 25 of *Biblioteca delle tradizioni popolari siciliane* (Palermo: A. Reber, 1913), p. 36.

18 Useful on Mediterranean concepts of honor and shame are: Jane Schneider, "Of Vigilance and Virgins: Honor, Shame and Access to Resources in Mediterranean Societies," *Ethnology* 10 (1971): 1–24, and the 1930s field work reported in Chapman, *Milocca*, pp. 72–80, 109–10. See also Rudolf Bell, *Fate and Honor, Family and Village* (Chicago: University of Chicago Press, 1979).

19 The best general source on the work camps of men around the world remains Robert Foerster, *The Italian Emigration of Our Times* (New York: Russell & Russell, 1968). See also vivid descriptions in Michael La Sorte, *La Merica: Images of Italian Greenhorn Experience* (Philadelphia, Pa.: Temple University Press, 1985), ch. 3; Robert F. Harney, "The Padrone System and Sojourners in the Canadian North, 1885–1920," in George E. Pozzetta (ed.), *Pane e Lavoro: The Italian American Working Class* (Toronto: Multicultural History Society of Ontario, 1980), pp. 119–40.

20 Harney, "Men Without Women," p. 88 quoting Peter Roberts, *The New Immigration*, p. 115.

21 La Sorte, *La Merica*, ch. 4; see also Robert F. Harney, "Boarding and Belonging: Thoughts on Sojourner Institutions," *Urban History Review* 2 (1978): 8–37. Ultimately 90 percent of Italians who remained in the U.S. settled in cities; in Argentina the figure was 70 percent. See Samuel L. Baily, *Immigrants in the Lands of Promise: Italians in Buenos Aires and New York City, 1870–1914* (Ithaca, N.Y.: Cornell University Press, 1999).

22 Donna Gabaccia, "The 'Yellow Peril' and the 'Chinese of Europe': Global Perspectives on Race and Labour, 1815–1930," in Jan and Leo Lucassen (eds), *Migrations, Migration History, History: Old Paradigms and New Perspectives* (New York: Berg and International Institute for Social History, 1997), pp. 177–96.

23 Quoted in Harney, "Men without Women," p. 52.

24 René Del Fabbro, *Transalpini: Italienische Arbeitswanderung nach Süddeutschland im Kaiserreich 1870–1918* (Osnabrück: Universitätsverlag Rasch, 1996), p. 217.

25 La Sorte, *La Merica*, p. 138.

26 Foerster, *The Italian Emigration of Our Times*, p. 147.

27 Quoted in Jerre Mangione and Ben Morreale, *La Storia; Five Centuries of the Italian American Experience* (New York: HarperCollins, 1992), p. 86.

28 La Sorte, *La Merica*, p. 99.

29 Harney, "Men Without Women," p. 89.

30 La Sorte, *La Merica*, p. 61.

31 Caroline Brettell, *The Men Who Migrate and the Women Who Wait: Population and History in a Portuguese Parish* (Princeton: University of Princeton Press, 1986).

32 Linda S. Reeder, "Widows in White: Sicilian Women and Mass Migration, 1880–1930," unpublished Ph.D. thesis, Rutgers University, 1995. A useful study of women who waited in Italy's North is Patrizia Audenino, "Le custodi della montagna: Donne e migrazioni stagionali in una comunità alpina," *Annali dell'Istituto Alcide Cervi* 12 (1990): 265–88. See also Vito Teti, "Noti sui comportamenti delle donne sole degli 'americani' durante la prima emigrazione in Calabria," *Studi*

emigrazione 24 (1987): 13–46; Paola Corti, "Sociétés sans hommes et intégrations des femmes à l'étranger; Le cas de l'Italie," *Revue européenne des migrations internationales* 9, 2 (1993): 113–28.

33 *Inchiesta parlamentare sulle condizioni dei contadini* (Rome: Tip. Naz. di Giovanni Bertero, 1910), vol. 5, p. 591.

34 Massimo Livi-Bacci, *A History of Italian Fertility during the Last Two Centuries* (Princeton, N.J.: Princeton University Press, 1977), pp. 69–70.

35 Donna Gabaccia, *Militants and Migrants: Rural Sicilians Become Italian Workers* (New Brunswick: Rutgers University Press, 1998), p. 87.

36 *Inchiesta parlamentare sulle condizioni dei contadini*, vol. 5, p. 591.

37 Anne Cornelisen, *Women of the Shadows*, quoted in Harney, "Men Without Women," p. 86.

38 Besides Elda Zappi, *If Eight Hours Seem Too Few: Mobilization of Women Workers in the Italian Rice Fields* (Albany: State University of New York Press, 1991), and Del Fabbro, *Transalpini*, pp. 264–5, see *Atti della Giunta parlamentare per l'inchiesta agraria*, vol. 4, fasc. 1, pp. 8–17 (on the Veneto) and vol. 8, tome 1, fasc. 2, pp. 628–32 (on various regions of Piedmont).

39 *Atti della Giunta parlamentare per l'inchiesta agraria*, vol. 8, tome 2, fasc. I, p. 126.

40 Ibid., vol. 4, p. 259.

41 Emma Ciccotosto and Michal Bosworth, *Emma, A Translated Life* (Fremantle, Australia: Fremantle Press, 1990), pp. 22–5.

42 *Inchiesta parlamentare sulle condizioni dei contadini*, vol. 4, p. 257.

43 Gabaccia, *From Sicily to Elizabeth Street*, ch. 3; see Reeder, "Widows in White."

44 *Atti della Giunta parlamentare per l'inchiesta agraria*, vol. 9, p. 327 and vol. 11, tome 1, p. 786.

45 Gabaccia, *From Sicily to Elizabeth Street*, p. 124.

46 Pitrè, *Proverbi siciliani*, vol. 2, p. 92.

47 Besides Ramella, *Terra e telai*, see Louise A. Tilly, *Politics and Class in Milan, 1881–1901* (New York: Oxford University Press, 1992), pp. 71–2.

48 Gabaccia, "In the Shadows of the Periphery."

49 *Atti della Giunta parlamentare per l'inchiesta agraria*, vol. 4, fasc. 1, p. 38.

50 Reeder, "Women in White"; Linda S. Reeder, "When the Men Left Sutera: Sicilian Women and Mass Migration 1880–1920," in Donna Gabaccia and Franca Iacovetta (eds), *Foreign, Female and Fighting Back: Women and the Italian Diaspora* (forthcoming).

51 Nina Glick Schiller, Linda Basch, and Cristina Blanc-Szanton, *Nations Unbound, Transnational Projects, Postcolonial Predicaments, and Deterritorialized Nation-States* (London: Gordon & Breach, 1994), p. 240.

52 On wages, see Foerster, *The Italian Emigration of Our Times*, pp. 143–4, 166, 200, 378–9.

53 *Atti della Giunta parlamentare per l'inchiesta agraria*, vol. 12, fasc. 1, p. 486.

54 Michael Burawoy, "The Functions and Reproduction of Migrant Labor: Comparative Material from Southern Africa and the United States," *American Journal of Sociology* 81 (March 1976): 1076–87. There is a sizeable and useful scholarly literature, inspired by feminist socialist theorists, on the reproduction of labor power. See Kate Young, Carol Wolkowitz, and Roslyn McCullagh (eds), *Of Marriage and the Market:*

Women's Subordination in International Perspective (London: CSE Books, 1981); Annette Kuhn and AnneMarie Wolpe, *Feminism and Materialism: Women and Modes of Production* (London: Routledge and Kegan Paul, 1978); Claude Meillassoux, *Maidens, Meal, and Money: Capitalism and the Domestic Economy* (Cambridge: Cambridge University Press, 1981). See also Joan Smith, Immanuel Wallerstein, and Hans-Dieter Evers (eds), *Households and the World Economy* (Beverly Hills, Calif.: Sage Publications, 1984).

55 *Atti della Giunta parlamentare per l'inchiesta agraria*, vol. 7, fasc. 2, p. 347.

56 *Inchiesta parlamentare sulle condizioni dei contadini*, vol. 2, p. 258.

57 On remittances, see *Annuario statistico della emigrazione italiana* (Rome: Commissariato dell'Emigrazione, 1927), pp. 1637–45; Dino Cinel, *The National Integration of Italian Return Migration, 1870–1929* (Cambridge: Cambridge University Press, 1991), pp. 141–4.

58 *Inchiesta parlamentare sulle condizioni dei contadini*, vol. 4: *Caserta*.

59 Robert Coit Chapin, *The Standard of Living Among Workingmen's Families in New York City* (New York: Charities Publication Committee, 1909); see also Louise Bolard More, *Wage Earners' Budgets*, Greenwich House Series of Social Studies, 1 (New York: Henry Holt, 1907).

60 Baily, *Immigrants in the Lands of Promise*, pp. 110–13.

61 Donna Gabaccia, "Women of the Mass Migrations: From Minority to Majority, 1820–1930," in Dirk Hoerder and Leslie Page Moch (eds), *European Migrants: Global and Local Perspectives* (Boston, Mass.: Northeastern University Press, 1996), p. 3.

62 John Davis, *Land and Family in Pisticci*, London School of Economics Monographs on Social Anthropology, 48 (New York: Humanities Press, 1973). Reeder, "Widows in White," pp. 219–20, shows that in Sutera, parents who had once dowered or, on their deaths, passed on houses to daughters, began giving them instead to sons during the migrations. See also Andreina De Clementi, "Proprietà, parentela e sistema successorio nell'emigrazione meridionale del primo Novecento," *Passato e presente* 28 (January–April 1993): 163–76.

63 On dowering customs, see Pitrè, *La famiglia, la casa, la vita*, pp. 34, 89; Giuseppe Pitrè, *Usi e costumi, credenze e pregiudizi del popolo siciliano*, vols. 14–17 of *Biblioteca delle tradizioni popolari siciliane* (Palermo: Libreria L. Pedone Lauriel di Carlo Clausen, 1889), vol. 15, pp. 9–12.

64 Dino Cinel, *From Italy to San Francisco* (Stanford, Calif.: Stanford University Press, 1982), p. 13.

65 On class relations in rural Italy, see: Renzo Del Carria, *Proletaria senza rivoluzione*, 4 vols., (Rome: Sanelli, 1975); Franco Della Peruta, *Società e classi popolari nell'Italia dell'Ottocento* (Syracuse: Ediprint, 1985).

66 Besides Cinel, *National Integration of Italian Return Migration*, and Betty Boyd Caroli, *Italian Repatriation from the United States, 1900–1914* (New York: Center for Migration Studies, 1973), see the work of Francesco Paolo Cerase: "Su una tipologia di emigrati ritornati: Il ritorno di investimento," *Studi emigrazione* 4, 10 (1967): 327–44; "Expectations and Reality: A Case Study of Return Migration from the United States to Southern Italy," *International Migration Review* 8 (1974): 245–62; *L'emigrazione di ritorno: Innovazione o reazione? L'esperienza dell'emigrazione di ritorno dagli Stati Uniti d'America* (Rome: Istituto Gini, 1971); George

Gilkey, "The United States and Italy: Migration and Repatriation," *Journal of Developing Areas* 2 (1967): 23–35; Joseph Lopreato, *Peasants No More* (San Francisco, Calif.: Chandler, 1967).

67 Cinel, *National Integration of Italian Return Migration*, pp. 164–8.

68 The best source here remains the six volumes of the *Inchiesta parlamentare sulle condizioni dei contadini*.

69 Antonio Papa, "Guerra e terra," *Studi storici* 10 (1969): 3–45.

70 Davis, *Land and Family in Pisticci*; Gabaccia, *From Sicily to Elizabeth Street*, pp. 32–3.

71 *Inchiesta parlamentare sulle condizioni dei contadini*, vol. 5, p. 504.

72 Ibid., vol. 2, pp. 186–7.

73 Quotes in Foerster, *The Italian Emigration of Our Times*, p. 105.

74 Cinel, *National Integration of Italian Return Migration*, ch. 7, does not even mention shops and businesses as an important outlet for returners' investments. See also Gabaccia, *Militants and Migrants*, pp. 158–60. Comparisons to contempory returners are sometimes useful. See Russell King, *Il ritorno in patria: Return Migration to Italy in Historical Perspective*, Occasional publications (new series), 23 (Durham: University of Durham Department of Geography, 1988); Peter Kammerer, "Ricerca sui rientri a Monopoli," *Inchiesta* 13, 62 (1983): 59–61 or Peter Kammerer, *Arbeit gibt's immer* (Frankfurt: Campus Verlag, 1987). For a theoretical discussion, see Saskia Sassen-Koob, "The International Circulation of Resources and Development: The Case of Migrant Labour," *Development and Change* 9 (October 1978): 509–45.

75 *Atti della Giunta parlamentare per l'inchiesta agraria*, vol. 5, tome 2, p. 154.

76 Ibid., vol. 1, p. 236.

77 For an overview, Stefano Somogyi, "L'alimentazione nell'Italia unita," in *Storia d'Italia*, vol. 5: *I documenti* (Turin: Einaudi, 1973), pp. 841–87. On pellagra, see Giorgio Porisini, *Agricoltura, alimentazione e condizioni sanitarie: Prime ricerche sulla pellagra in Italia dal 1880 al 1940* (Geneva: Librairie Droz, 1979).

78 *Atti della Giunta parlamentare per l'inchiesta agraria* contains detailed notes on the diet of peasants in all of Italy's regions in the 1870s and 1880s.

79 *Inchiesta parlamentare sulle condizioni dei contadini*, vol. 2, p. 159.

80 Piero Bevilacqua, "Emigrazione transoceanica e mutamenti dell'alimentazione contadina calabrese fra Otto e Novecento," *Quaderni storici* 47, 2 (1981): 520–55; Vito Teti, "La cucina calabrese è un invenzione americana," *I viaggi di Erodoto* 12 (1991): 58–73.

81 Paola Corti, "Il cibo dell'emigrante," *Il Risorgimento* 44, 2 (1992): 363–78.

82 See, e.g., *Inchiesta parlamentare sulle condizioni dei contadini*, vol. 4: *Salerno.*

83 *Atti della Giunta parlamentare per l'inchiesta agraria*, vol. 31, fasc. 2, appendices, p. 113; see also vol. 12, fasc. 4, for the region around Bari and Lecce where serenading was called "potagalli."

84 *Inchiesta parlamentare sulle condizioni dei contadini*, vol. 2, p. 271.

85 Ibid., p. 251.

86 Ibid., vol. 3, p. 545.

87 Gabaccia, *Militants and Migrants*, Table 4.2.

88 Mangione and Morreale, *La Storia*, p. 72.

89 Dino Cinel, "Land Tenure Systems, Return Migration and Militancy in Italy," *Journal of Ethnic Studies* 12 (1984): 55–76.

90 Del Fabbro, *Transalpini*, pp. 266–7.

91 See Chapter 6 for the ideologically freighted use of the term colony by Italian speakers.

92 The case study of Italian migrants in particular cities and neighborhoods is a far more popular methodology in English-speaking North America than elsewhere in the world. A good starting place is Robert F. Harney and J. Vincenza Scarpaci (eds), *Little Italies in North America* (Toronto: Multicultural History Society of Ontario, 1981). For the U.S., see works cited in Chapter 3, notes 77 and 79–83. In addition, on Chicago, Rudolph J. Vecoli, "The Formation of Chicago's 'Little Italies,'" *Journal of American Ethnic History* 1 (Spring 1983): 5–20; on St. Louis, Gary Mormino, *Immigrants on the Hill: Italian-Americans in St. Louis 1882–1982* (Urbana: University of Illinois Press, 1986). On San Francisco, Micaela Di Leonardo, *The Varieties Of Ethnic Experience* (Ithaca, N.Y.: Cornell University Press, 1984). On Philadelphia, Richard Juliani, *The Social Organization of Immigration: The Italians of Philadelphia* (New York: Arno, 1981). For Canada, see Robert F. Harney, "Chiaroscuro: Italians in Toronto, 1885–1915," *Polyphony* 6, 1 (1984): 44–9; John Zucchi, "Italian Hometown Settlements and the Development of an Italian Community in Toronto, 1875–1935," in Robert F. Harney (ed.), *Gathering Place: Peoples and Neighbourhoods of Toronto* (Toronto: Multicultural History Society of Ontario, 1985), pp. 121–46; Bruno Ramirez, *Les Premiers Italiens de Montréal: L'origine de la Petite Italie du Québec* (Montreal: Boreal Express, 1984). The most extensively studied Italian settlement in Argentina, besides Buenos Aires, is La Boca (just outside that city). See Fernando J. Devoto, "The Origins of an Italian Neighborhood in Buenos Aires in the Mid XIX Century," *Journal of European Economic History* 18, 1 (Spring 1989): 37–64; Dora Barrancos, "Vita materiale e battaglia ideologica nel quartiere della Boca (1880–1930)," in Gianfausto Rosoli (ed.), *Identità degli italiani in Argentina* (Rome: Stadium, 1993), pp. 167–204. For São Paulo, M.F. Alvim Zuleika and Jose Sachetta Ramos, "Italianos en São Paulo; Dimensiones de la italianidad en el estado de São Paulo en 1920," *Estudios migratorios Latinoamericanos* 29 (April 1995): 113–28; Tania Regina de Luca, "Inmigracion, mutualismo e identidad: São Paulo (1890–1935)," *Estudios migratorios Latinoamericanos* 29 (April 1995): 191–208. A good collection of studies of local Italian settlements in France are in Pierre Milza (ed.), *Les Italiens en France de 1914 à 1940* (Rome: L'École Française de Rome, 1986). French scholars have more often preferred regional studies. See, e.g., Pierre Milza, Antonio Bechelloni, and Michel Dreyfus (eds), *L'Intégration italienne en France: Un siècle de présence italienne dans trois regions françaises: 1880–1980* (Brussels: Complexe, 1995); Paola Corti and Ralph Schor (eds), "L'esodo frontaliero: Gli italiani nella Francia meridionale," special issue, *Recherches Régionales* (1995); Anna Maria Faidutti-Rudolph, *L'Immigration italienne dans le sud-est de la France* (Gap: Ophrys, 1964).

93 Family budgets and the internal dynamics of migrant families are little studied outside the U.S. This reflects two somewhat unique developments in U.S. historiography — the predominance of social histories of workers in the 1970s and 1980s and the proliferation of

studies of working-class women, including immigrant women, not just as workers but as consumers. Good starting places are: Susan Porter Benson, "Living on the Margin: Working-Class Marriages and Family Survival Strategies in the United States, 1919–1941," in Victoria de Grazia (ed.), *The Sex of Things: Gender and Consumption in Historical Perspective* (Berkeley: University of California Press, 1996); and Lizabeth Cohen, "The Class Experience of Mass Consumption: Workers in Interwar America," in Richard Wightman Fox and T.J. Jackson Lears (eds), *The Power of Culture: Critical Essays in American History* (Chicago: University of Chicago Press, 1993). For Argentina, besides Baily, *Immigrants in the Promised Lands*, see Maria Cristina Cacopardo and José Luis Moreno, *La familia italiana y meridional en la emigracion a la Argentina* (Naples: Scientifiche Italiane, 1994).

94 *Inchiesta parlamentare sulle condizioni dei contadini*, vol. 4, p. 614, vol. 5, p. 711.

95 Baily, *Immigrants in the Lands of Promise*, Table 7.7; Miriam Cohen, *Workshop to Office: Two Generations of Italian Women in New York City, 1900–1950* (Ithaca, N.Y.: Cornell University Press, 1993); Elizabeth Ewen, *Immigrant Women in the Land of Dollars; Life and Culture on the Lower East Side, 1890–1925* (New York: Monthly Review Press, 1985), ch. 6; Nancy L. Green, *Ready-to-Wear; Ready-to-Work: A Century of Industry and Immigrants in Paris and New York* (Durham, N.C.: Duke University Press, 1997), ch. 6.

96 For a contemporary's report, see Mabel Hurd Willett, *The Employment of Women in the Clothing Trade*, Columbia University Studies in History, Economics and Public Law, 16 (New York: Columbia University Press, 1902); for a recent assessment, Cynthia R. Daniels, "Between Home and Factory: Homeworkers and the State," in Eileen Boris and Cynthia R. Daniels (eds), *Homework: Historical and Contemporary Perspectives on Paid Labor at Home* (Urbana: University of Illinois Press, 1989).

97 Lillian Betts, "Italian Peasants in a New Law Tenement," *Harper's Bazaar* 38 (1904): 804; Gabaccia, *From Sicily to Elizabeth Street*, p. 64; see also Thomas Kessner and Betty Boyd Caroli, "New Immigrant Women at Work: Italians and Jews in New York City, 1880–1905," *Journal of Ethnic Studies* 5 (1978): 19–32. Louise C. Odencrantz, *Italian Women in Industry, A Study of Conditions in New York City* (New York: Russell Sage Foundation, 1919), pp. 16 and 21, also found 24 percent of married Italian women working outside the home.

98 Foerster, *Italian Emigration of Our Times*, pp. 144–7.

99 Baily, *Immigrants in the Lands of Promise,* Tables 7.5 and 7.6. See also Harney, "Boarding and Belonging." For general patterns in the U.S., see John Modell and Tamara K. Hareven, "Urbanization and the Malleable Household: An Examination of Boarding and Lodging in American Families," *Journal of Marriage and the Family* 35 (1973): 467–79.

100 Chapin, *The Standard of Living Among Workingmen's Families*, pp. 245–50. See also State of New York, *Factory Commission Report* (1915), vol. 4, pp. 1608–9, 1619–20, 1668.

101 Baily, *Immigrants in the Lands of Promise*, Tables 5.5 and 5.6.

102 Gabaccia, *From Sicily to Elizabeth Street*, p. 92.

103 On Argentina, see Paola Corti, "Circuiti migratori e consuetudini alimentari nella rappresentazione autobiografica degli emigranti piemontesi," in A. Capatti, A. De Bernardi, and A. Varni, *L'alimentazione nella storia dell'Italia contemporanea, Storia d'Italia, Annali* (Turin: Einaudi, 1997); Peppino Ortoleva, "La tradizione e l'abbondanza; Riflessioni sulla cucina degli italiani d'America," *Altreitalie* 7 (1992): 31–52; Arnd Schneider, "L'etnicità, il cambiamento dei paradigmi e le variazioni nel consumo di cibi tra gli italiani a Buenos Aires," *Altreitalie* 7 (January–June 1992): 71–83. On the U.S., more generally, Donna R. Gabaccia, *We Are What We Eat: Ethnic Food and the Making of Americans* (Cambridge, Mass.: Harvard University Press, 1998), pp. 51–5; Donna Gabaccia, "Italian-American Cookbooks: From Oral to Print Culture," and Carol Helstosky, "The Tradition of Invention: Reading History through 'La Cucina Casareccia Napoletana,'" *Italian Americana* 16, 1 (Winter 1998): 7–23.

104 Gabaccia, *From Sicily to Elizabeth Street*, pp. 101–2.

105 Direct quotes are from Judith E. Smith, *Family Connections: A History of Italian and Jewish Immigrant Lives in Providence, Rhode Island, 1900–1940* (Albany: State University of New York Press, 1985), p. 75.

106 Kathy Peiss, *Cheap Amusements: Working Women and Leisure in Turn-of-the-Century New York* (Philadelphia, Pa.: Temple University Press, 1986); Ewen, *Immigrant Women in the Land of Dollars*, pp. 232–3.

107 Pitrè, *La famiglia, la casa, la vita*, pp. 34, 89.

108 Gabaccia, *From Sicily to Elizabeth Street*, p. 97, n. 95.

109 Corinne Azen Krause, *Grandmothers, Mothers, and Daughters; Oral Histories of Three Generations of Ethnic American Women* (Boston, Mass.: Twayne Publishers, 1991), p. 22.

110 Baily, *Immigrants in the Lands of Promise*, ch. 6. On partner households, see Gabaccia, *From Sicily to Elizabeth Street*, pp. 75–6.

111 Gabaccia, *We Are What We Eat*, pp. 52–3.

112 Baily, *Immigrants in the Lands of Promise*, Table 5.7; see also Samuel L. Baily, "Patrones de residencia de los italianos en Buenos Aires y Nueva York: 1880–1914," *Estudios Migratorios Latinoamericanos* 1 (December 1985): 8–47.

113 Baily, *Immigrants in the Lands of Promise*, Tables 6.1 and 6.2. On the Cinisi, Robert E. Park and Herbert A. Miller, *Old World Traits Transplanted* (New York: Harper and Brothers, 1925), pp. 146–58; Pitrè, *La famiglia, la casa, la vita*, p. 41.

5 NATIONALISM AND INTERNATIONALISM IN ITALY'S PROLETARIAN DIASPORAS, 1870–1914

1 Hermia Oliver, *The International Anarchist Movement in Late Victorian London* (London: Croom Helm, 1983), Preface.

2 I use the term labor internationalism to distinguish the internationalism of the anarchist, and of the First, Second, and Third Internationals from the internationalist republicanism of the era of Garibaldi and Mazzini.

3 For an early effort, see Giuseppe Prato, *Il protezionismo operaio; L'esclusione del lavoro straniero* (Turin: Collegio degli Artigianelli, 1910).

4 For an introduction to these issues, see Marcel van der Linden, "The Rise and Fall of the First International," and Theo van Tijn, "Nationalism and the Socialist Workers Movement," in Frits van Holthoon and Marcel van der Linden (eds), *Internationalism in the Labour Movement, 1830–1940* (Leiden: E.J. Brill, 1988).

5 The early insights of Ernesto Ragionieri, "Italiani all'estero ed emigrazione di lavoratori italiani: Un tema di storia del movimento operaio," *Belfagor, Rassegna di varia umanità* 17, 6 (1962): 640–69, and empirical compilations in Bruno Bezza, *Gli italiani fuori d'Italia, Gli emigrati italiani nei movimenti operai dei paesi d'adozione 1880–1940* (Milan: Franco Angeli, 1983), have been built on in recent studies: Donna R. Gabaccia, "Worker Internationalism and Italian Labor Migration, 1870–1914," *International Labor and Working-Class History* 45 (Spring 1994): 63–70; Donna Gabaccia and Fraser Ottanelli, "Diaspora or International Proletariat? Italian Labor, Labor Migration, and the Making of Multiethnic States, 1815–1939," *Diaspora* 6, 1 (Spring 1997): 61–84.

6 I borrow the term "other Italies" from the periodical published by the Giovanni Agnelli Foundation, *Altreitalie* (1988–94).

7 Useful introductions to the concept of leadership in migrant communities are: Victor R. Green, *American Immigrant Leaders, 1800–1910: Marginality and Identity* (Baltimore, Md.: Johns Hopkins University Press, 1987); Victor R. Green, "Becoming American: The Role of Ethnic Leaders; Swedes, Poles, Italians, and Jews," in Melvin G. Holli and Peter Jones (eds), *The Ethnic Frontier: Essays in the History of Group Survival in Chicago and the Mid-West* (Grand Rapids, Mich.: Eerdmans, 1977); John Higham (ed.), *Ethnic Leadership in America* (Baltimore, Md.: Johns Hopkins University Press, 1978).

8 On the later exploits of the garibaldini, see the reverent account of Antonio Bandini Buti, *Una epopea sconosciuta: È il contributo di fede e di sangue dato dai volontari italiani per la causa degli altri popoli dal Trocadero e Missolunghi alle Argonne e Bligny (1818–1918)* (Milan: Ceschina, 1967). On Garibaldi as a socialist, see Letterio Briguglio, *Garibaldi e il socialismo* (Rome: Laterza, 1984).

9 Besides Bianca Montale, *Mazzini e le origini del movimento operaio: Appunti di storia del Risorgimento* (Genoa: Tilgher, 1976), see the general studies: Lia Gheza Fabbri, *Solidarismo in Italia fra XIX e XX secolo: Le società di mutuo soccorso e le casse rurali* (Turin: G. Giappichelli, 1996), and Corrado Perna, *Breve storia del sindacato: Dalle società di mutuo soccorso al sindacato dei consigli* (Bari: De Donato, 1978). For developments in the South, see: Italy, Direzione generale della statistica, *Elenco delle società di mutuo soccorso* (1886–96); Giuseppe Cantarella, *Società operaie di mutuo soccorso e società cooperative nella Provincia di Reggio Calabria fra il 1858 ed il 1908* (Cosenza: Progetto 2000, 1989).

10 On the origins of Catholic activism among peasants and laborers, see: *La "Rerum novarum" e il movimento cattolico italiano* (Brescia: Morcelliana, 1995); Pietro Borzomati, *Movimento cattolico e Mezzogiorno* (Rome: La Goliardica, 1982); Francesco Renda, *Socialisti e cattolici in Sicilia, 1900–1904; Le lotte agrarie* (Caltanissetta: Salvatore Sciascia, 1972).

11 Gino Cerrito, "Saverio Friscia nel primo periodo di attività dell'Inter-

nazionale in Sicilia," *Movimento operaio* 5 (1953): 464–75; Nazario Galassi, *Vita di Andrea Costa* (Milan: Feltrinelli, 1989).

12 See biography in Franco Andreucci and Tommaso Detti (eds), *Il movimento operaio italiano, Dizioniario biografico, 1853–1943* (Rome: Riuniti, 1975–8), vol. 1. Although a biography of Bignami also appears in Michele Rosi, *Dizionario del Risorgimento nazionale; Dalle origini a Roma Capitale; Fatti e persone* (Milan: Casa Dottor Francesco Vallardi, 1930–7), it ignores his conversion to socialism.

13 My descriptions of the characteristics and migrations of Italy's labor exiles are based on an analysis of 411 leaders and activists in Andreucci and Detti, *Il movimento operaio italiano.*

14 Older English-language studies especially emphasize the peculiarities of the Italian labor movement: Joseph La Palombara, *The Italian Labor Movement: Problems and Prospects* (Ithaca, N.Y.: Cornell University Press, 1957); Maurice F. Neufeld, *Italy: School for Awakening Countries, The Italian Labor Movement in its Political, Social, and Economic Setting from 1800 to 1960* (Ithaca, N.Y.: Cornell University Press, 1961); Daniel L. Horowitz, *The Italian Labor Movement* (Cambridge, Mass.: Harvard University Press, 1963).

15 On the labor movement's anarchist birth, see: Adriana Dadà, *L'anarchismo in Italia fra movimento e partito: Storia e documenti dell'anarchismo italiana* (Milan: Teti, 1984), p. 23; Franco Damiani, *Bakunin nell'Italia postunitaria, 1864–1867; Anticlericalismo, democrazia, questioni operaia e contadina negli anni del soggiorno italiano di Bakunin* (Milan: Jaca Book, 1977), pt. 1.

16 On Italy's anarchist movement, see Nunzio Pernicone, *Italian Anarchism, 1864–1892* (Princeton, N.J.: Princeton University Press, 1993); Lilian Faenza (ed.), *Anarchismo e socialismo in Italia, 1872–1892* (Rome: Riuniti, 1973); Eva Civolani, *L'anarchismo dopo la Comune: I casi italiano e spagnolo* (Milan: Franco Angeli, 1981).

17 Andreucci and Detti, *Il movimento operaio italiano*, vol. 3.

18 See Jole Calapso, *Donne ribelli: Un secolo di lotte femminili in Sicilia* (Palermo: S.F. Flaccovio, 1980).

19 On Malatesta and anarcho-syndicalism, see Enzo Santarelli, *Il socialismo anarchico in Italia* (Milan: Feltrinelli, 1973); Stefano Arcangeli, *Errico Malatesta e il comunismo anarchico italiano* (Milan: Jaca Book, 1972). For propaganda of the deed and Galleani's anarchism, see Pier Carlo Masini, *Storia degli anarchici italiani nell'epoca degli attentati* (Milan: Rizzoli, 1981); Ugo Fedeli, *Luigi Galleani: Quarant'anni di lotte rivoluzionarie (1891–1931)* (Cesena: L'Antistato, 1956).

20 Charles, Louise, and Richard Tilly, *The Rebellious Century, 1830–1930* (Cambridge, Mass.: Harvard University Press, 1975), p. 94.

21 Maria Grazia Meriggi, *Il Partito Operaio Italiano: Attività rivendicativa, formazione e cultura dei militanti in Lombardia (1880–1890)* (Milan: Franco Angeli, 1985). Costa later became a socialist, and served as a socialist deputy.

22 Besides Elda Zappi, *If Eight Hours Seem Too Few: Mobilization of Women Workers in the Italian Rice Fields* (Albany: State University of New York Press, 1991), and Calapso, *Donne ribelli*, see: Aurelia Camparini, "Lotte sociali e organizzazioni femminili, 1880–1926," in Aldo Agosti

and Gian Mario Bravo (eds), *Storia del movimento operaio; Del socialismo e delle lotte sociali in Piemonte*, 2 vols. (Bari: De Donato, 1979).

23 Ernesto Ragionieri, *Socialdemocrazia tedesca e socialisti italiani, 1875–1895* (Milan: Feltrinelli, 1961).

24 Corrado Dollo, *I fasci siciliani*, 2 vols. (Bari: De Donato, 1975, 1976); Francesco Renda, *I fasci siciliani, 1892–1894* (Turin: G. Einaudi, 1977), appendix 2; Giuseppe Casarrubea, *I fasci contadini e le origini delle sezioni socialiste della Provincia di Palermo*, 2 vols. (Palermo: S.F. Flaccovio, 1978).

25 Francesco Bogliari, *Il movimento contadino in Italia dall'unità al fascismo* (Turin: Loescher, 1980); Valeria Evangelisti and Salvatore Sechi, *Il galletto rosso: Precariato e conflitto di classe in Emilia-Romagna, 1880–1980* (Venice: Marsilio, 1982).

26 Besides general studies like Alberto Bonifazi and Gianni Salvarani, *Dalla parte dei lavoratori, Storia del movimento sindacale italiano*, 2 vols. (Milan: Franco Angeli, 1976), see Maurizio Antonioli and Bruno Bezza (eds), *La FIOM dalle origini al fascismo, 1901–1924* (Bari: De Donato, 1978); Enrico Finzi, *Alle origini del movimento sindacale: I ferrovieri* (Bologna: Il Mulino, 1975).

27 Louise A. Tilly, "I fatti di maggio: The Working Class of Milan and the Rebellion of 1898," in Robert J. Bezucha (ed.), *Modern European Social History* (Lexington, Mass.: D.C. Heath, 1972), pp. 124–60; Louise Tilly, *Politics and Class in Milan 1881–1901* (New York: Oxford University Press, 1992); Volker Hünecke, *Arbeiterschaft und industrielle Revolution in Mailand: 1859–1892: Zur Entstehungsgeschichte der italienischen Industrie and Arbeiterbewegung* (Göttingen: Vandenhoeck and Ruprecht, 1978).

28 On the "battle of the tendencies" within the Socialist Party, see Ernesto Ragionieri, *Il movimento socialista in Italia (1850–1922)* (Milan: Teti, 1976).

29 Useful for these years is Giorgio Galli, *Storia del socialismo italiano* (Bari: Laterza, 1980); English readers can consult Alexander De Grand, *The Italian Left in the Twentieth Century* (Bloomington: Indiana University Press, 1989).

30 See: Paolo Pettra, *Ideologie costituzionali della sinistra italiana (1892–1974)* (Rome: Savelli, 1975); Salvatore Onufrio, *Socialismo e marxismo nella "Critica Sociale," 1892–1912* (Palermo: S.F. Flaccovio, 1980).

31 Donna Gabaccia, *Militants and Migrants: Rural Sicilians Become Italian Workers* (New Brunswick: Rutgers University Press, 1988), pp. 58–60.

32 Patrizia Sione, "Storia delle migrazioni e storia del movimento operaio: Il caso dei tessili comaschi e biellesi nel New Jersey (Usa), 1880–1913," in Vanni Blengino, Emilia Franzina, and Adolfo Pepe (eds), *La riscoperta delle Americhe; Lavoratori e sindacato nell'emigrazione italiana in America Latina 1870–1970* (Milan: Teti Editore, 1993), pp. 277–90.

33 My analysis of 126 migrants with names starting with the letters "B" and "C" in the Casellario Politico Centrale, 1896–1914, Central Italian State Archives, Rome. See: Pier Carlo Masini, "Biografie di 'sovversivi' compilate dai prefetti del Regno d'Italia," *Rivista storica del socialismo* 6, 13–14 (1961): 573–604; Luigi Bruti Liberati, "Casellario politico centrale," *Studia nordamericana* 2, 1 (1985): 81–3; Emilio Franzina, "L'emigrazione schedata; Lavoratori sovversivi all'estero e meccanismi

di controllo poliziesco in Italia tra fine secolo e fascismo," in Bezza, *Gli italiani fuori d'Italia*, pp. 773–830.

34 Gabaccia, "Worker Internationalism and Italian Labor Migration, 1870–1914."

35 Gianfranco Cresciani, "L'integrazione dell'emigrazione italiana in Australia e la politica dell'trade unions: Dagli inizi del secolo al fascismo," in Bezza, *Gli italiani fuori d'Italia*, pp. 313, 319–21; Herbert Gutman, "The Buena Vista Affair, 1874–75," in Peter N. Stearns (ed.), *Workers in the Industrial Revolution: Recent Studies of Labor in the United States and Europe* (New Brunswick: Rutgers University Press, 1974); Robert Paris, "Le mouvement ouvrier français et l'immigration italienne (1893–1914)," in Bezza, *Gli italiani fuori d'Italia*, pp. 637–8; Enrico Serra, "L'emigrazione italiana in Francia durante il primo governo Crispi (1887–1891)," in Jean-Baptiste Duroselle and Enrico Serra (eds), *L'emigrazione italiana in Francia prima del 1914* (Milan: Franco Angeli, 1978), pp. 150–5 and 160–4; Marc Vuilleumier, "Les exilés en Suisse et le mouvement ouvrier socialiste (1871–1914)," in Maurizio Degl'Innocenti (ed.), *L'esilio nella storia del movimento operaio e l'emigrazione economica* (Manduria: Piero Lacaita, 1992), pp. 75–6; H. Looser, "Zwischen 'Tshinggenhass' und Rebellion: Der 'Italienerkrawall' von 1896," in *Lücken im Panorama; Einblicke in den Nachlass Zürichs* (Zurich: Geschichtsladen, 1986); Helmut Konrad, *Das Entstehen der Arbeiterklasse in Oberösterreich* (Vienna: Europaverlag, 1981), p. 368; Donald Avery, "Continental European Immigrant Workers in Canada, 1896–1910: From Stalwart Peasants to Radical Proletariat," *Canadian Review of Sociology and Anthropology* 12 (1975): 41.

36 Teodosio Vertone, "Antécédents et causes des événements d'Aigues-Mortes," in Duroselle and Serra, *L'emigrazione italiana in Francia*, pp. 116–34; E. Barnaba, "Aigues-Mortes: Una tragedia dell'emigrazione italiana in Francia," in Jean Charles Vegliante (ed.), *Gli italiani all'estero-3 bis: Autres passages* (Paris: Université de la Sorbonne, 1986), pp. 44–83.

37 Paris, "Le mouvement," p. 638; see also Van Tijn, "Nationalism and the Socialist Workers Movement."

38 Cresciani, "L'integrazione," pp. 307–12, 317–21. The quote is from William A. Douglass, *From Italy to Ingham: Italians in North Queensland* (St. Lucia: University of Queensland Press, 1995), p. 49.

39 Serra, "L'emigrazione italiana in Francia," pp. 47–9; Paris, "Le mouvement," pp. 638 and 644.

40 Vuilleumier, "Les exilés en Suisse," pp. 75–8.

41 Gwendolyn Mink, *Old Labor and New Immigrants in American Political Development* (Ithaca, N.Y.: Cornell University Press, 1986); Rudolph J. Vecoli, "Etnia, internazionalismo e protezionismo operaio: Gli immigrati italiani e i movimenti operai negli USA, 1880–1950," in Blengino *et al.*, *La riscoperta delle Americhe*, pp. 507–52; Catherine Collomp, "Les organisations ouvrières et la restriction de l'immigration aux États-Unis à la fin du dix-neuvième siècle," Marianne Debouzy (ed.), *In the Shadow of the Statue of Liberty: Immigrants, Workers and Citizens in the American Republic, 1880–1920* (St. Denis: Presses Universitaires de Vincennes, 1988) (bilingual Anglo-French edition).

42 On Canada see Bruno Ramirez, "Workers without a Cause: Italian Immigrant Labour in Montreal, 1880–1930," in Roberto Perin and Franc Sturino (eds), *Arrangiarsi: The Italian Immigration Experience in Canada* (Montreal: Guernica, 1992), pp. 119–34; Angelo Principe, "Note sul radicalismo tra gli italiani in Canada (1900–1915)," in Valeria Gennaro Lerda (ed.), *Movimento migratorio italiano: Canada e Stati Uniti* (Venice: Marsilio, 1984), pp. 147–56.

43 Good starting places on a large and growing literature are: Rudolph Vecoli, "The Italian Immigrants in the United States Labor Movement from 1880 to 1929," in Bezza, *Gli italiani fuori d'Italia*, pp. 257–306; Edwin Fenton, *Immigrants and Unions, a Case Study: Italians and American Labor, 1870–1920* (New York: Arno Press, 1975); Edwin Fenton, "Italians in the Labor Movement," *Pennsylvania History* 26 (1959): 33–143; Amelia Paparazzo, *Italiani del sud in America; Vita quotidiana, occupazione, lotte sindacali degli immigranti meridionali negli Stati Uniti (1880–1917)* (Milan: Franco Angeli, 1990).

44 Cresciani, "L'integrazione," pp. 321–2. Italian exiles numbered among the founders of the Australian Labor Party: Verity Burgman, *"In Our Time": Socialism and the Rise of Labor, 1885–1905* (Sydney: George Allen and Unwin, 1985), pp. 52, 101.

45 See biographies in Andreucci and Detti, *Il movimento operaio italiano*, vols. 3 and 5, and Franco Damiani, *Carlo Cafiero nella storia del primo socialismo italiano* (Milan: Jaca Book, 1974).

46 Nunzio Pernicone, "Luigi Galleani and Italian Anarchist Terrorism in the United States," *Studi emigrazione* 30 (September 1993): 469–88.

47 A useful comparison of differing labor movements is Sima Lieberman, *Labor Movements and Labor Thought: Spain, France, Germany and the United States* (New York: Praeger, 1986).

48 Ragionieri, "Italiani all' estero," pp. 648–51; Paris, "Le mouvement," pp. 636–63, 641–2; Pierre Milza, *Voyage en Ritalie* (Paris: Plon, 1993), p. 197.

49 Serra, "L'emigrazione italiana," p. 47.

50 Ragionieri, "Italiani all'estero," pp. 652–3.

51 Guido Pedroli, *Il socialismo nella Svizzera italiana (1880–1922)* (Milan: Feltrinelli, 1963), pp. 39–40.

52 Antonio Casali, "Emigrazione politica ed economica in Italia nell'età della Seconda Internazionale," in Degl'Innocenti, *L'esilio*, pp. 157–8.

53 Zeffiro Ciuffoletti, *L'emigrazione nella storia d'Italia, 1868/1975, Storia e documenti* (Florence: Vallecchi, 1978), doc. 42.

54 Enrico Decleva, *Etica del lavoro, socialismo, cultura popolare: Augusto Osimo e la Società Umanitaria* (Milan: Franco Angeli, 1985); Maurizio Punzo, "La Società Umanitaria e l'emigrazione, dagli inizi del secolo alla prima guerra mondiale," in Bezza, *Gli italiani fuori d'Italia*, pp. 119–44.

55 Hermann Schäfer, "L'immigrazione italiana nell'Impero Tedesco," in Bezza, *Gli italiani fuori d'Italia*, p. 756.

56 Mario Marcelletti, "Sindacati e problemi dell'emigrazione," in Franca Assante (ed.), *Movimento migratorio italiano dall'unità nazionale ai giorni nostri*, 2 vols. (Geneva: Librairie Droz, 1978), vol. 2, p. 3; Gianfausto Rosoli, "I sindacati di fronte all'emigrazione: Italia, Iugoslavia, Svizzera e Francia," in ibid., p. 111. See also Adolfo Pepe, "La Confederazione

Generale del Lavoro e l'emigrazione tra fine secolo e fascismo," in Blengino *et al.*, *La riscoperta delle Americhe*, pp. 15–34.

57 Zeffiro Ciuffoletti, "Il movimento sindacale italiano e l'emigrazione dalle origini al fascismo," in Bezza, *Gli italiani fuori d'Italia*, p. 214.

58 Susan Milner, "The International Labour Movement and the Limits of Internationalism: The International Secretariat of National Trade Union Centres, 1901–1913," *International Review of Social History* 33 (1988): 1–24.

59 Here the influence of German Social Democrats within the International may have been critical. See Martin Forberg, "Foreign Labour, the State and the Trade Unions in Imperial Germany, 1890–1918," in W.R. Lee and Eve Rosenhaft (eds), *The State and Social Change in Germany, 1880–1980* (New York: Berg/St. Martin's Press 1990); Claudie Weill, *L'Internationale et l'autre: Les relations inter-ethniques dans la IIe Internationale (Discussions et débats)* (Paris: de l'Arcantère, 1987), pp. 98–113.

60 See Chapter 1, note 2, on the interchangeability of "ethnic" and "national" in most European languages in the years around 1900.

61 Pierre Milza, "L'émigration italienne en France de 1870 à 1914," *Affari sociali internazionali* 5 (Fall–Winter 1977): 63–86; see also Pierre Milza, "L'intégration des italiens dans le mouvement ouvrier français à la fin de XIXe et au début du XXe siècle; le cas de la région Marseillaise," *Affari sociali internazionali* 5 (Fall–Winter 1977): 171–208; Paris, "Le mouvement," 663–78.

62 Nancy L. Green, *Ready-to-Wear; Ready-to-Work: A Century of Industry and Immigrants in Paris and New York* (Durham, N.C.: Duke University Press, 1977), p. 261.

63 Teobaldo Filesi, "Significato e portata della presenza italiana in Africa dalla fine del XVIII secolo ai nostri giorni," in Assante, *Movimento migratorio italiano*, vol. 2, pp. 387–425; Romain H. Rainero, "Gli emigrati italiani nel Nord Africa e la loro importanza nella nascità dei movimenti operai maghrebini (1880–1922)," in Bezza, *Gli italiani fuori d'Italia*, pp. 763–72.

64 Paris, "Le mouvement," p. 677.

65 Samuel L. Baily, "The Italians and the Development of Organized Labor in Argentina, Brazil, and the United States, 1880–1914," *Journal of Social History* 3, 2 (Winter 1969–70): 123–34; Samuel L. Baily, "The Italians and Organized Labor in the United States and Argentina," *International Migration Review* 1, 3 (Summer 1967): 55–66; Torcuato Di Tella, "Working Class Organization and Politics in Argentina," *Latin American Research Review* 16 (1982): 40–4; Osvaldo Bayer, "L'influenza dell'immigrazione italiana nel movimento anarchico argentino," in Bezza, *Gli italiani fuori d'Italia*, pp. 531–48; Ronaldo Munck, Ricardo Falcón, and Bernardo Galitelli, *Argentina: From Anarchism to Peronism, Workers, Unions and Politics, 1855–1985* (London: Zed Books, 1987); Ruth Thompson, "The Limitations of Ideology in the Early Argentine Labour Movement: Anarchism in the Trade Unions, 1890–1920," *Journal of Latin American Studies* 16 (May 1984): 81–99.

66 Newton Stadler De Souza, *O anarquismo da colonia Cecilia* (Rio de Janeiro: Civilizaçao, 1970); Marcello Zane, "'Turbulento anarquista e pesquisador

agronomico': Giovanni Rossi in Brasile dopo la Colonia Cecilia," *Studi bresciani* 6 (1992): 7–19.

67 Edgar Rodrigues, *Os anarquistas: Trabalhadores italianos no Brasil* (São Paulo: Global, 1984); Michael M. Hall and Paulo Sergio Pinheiro, "Immigrazione e movimento operaio in Brasile: Un'interpretazione," in José Luiz Del Roio (ed.), *Lavoratori in Brasile; Immigrazione e industrializzazione nello stato di San Paolo* (Milan: Franco Angeli, 1981), pp. 35–48; Michael Hall, "Immigration and the Early São Paulo Working Class," *Jahrbuch für Staat, Wirtschaft und Gesellschaft Lateinamerikas* 2 (1975): 393–407; Sheldon Maram, "The Immigrant and the Brazilian Labor Movement," in Dauril Alden and Warren Dean (eds), *Essays Concerning the Socioeconomic History of Brazil and Portuguese India* (Gainesville: University of Florida Press, 1977), pp. 178–210; Sheldon L. Maram, "Labor and the Left in Brazil, 1890–1921," *Hispanic American Historical Review* 57 (May 1977): 254–72; Jose Antonio Segatto, *A formação da classe operária no Brasil* (Porto Alegre: Mercado Aberto, 1987).

68 Anna Maria Martellone, "Per una storia della sinistra italiana negli Stati Uniti: Riformismo e sindacalismo, 1880–1922," in Assante, *Movimento migratorio italiano*, vol. 2, pp. 181–95. On language locals in the Socialist Party, Gary Marks and Matthew Burbank, "Immigrant Support for the American Socialist Party, 1912 and 1920," *Social Science History* 14 (Summer 1990): 175–202; Elisabetta Vezzosi, *Il socialismo indifferente: Immigrati italiani e Socialist Party negli Stati Uniti del primo Novecento* (Rome: Ed. Lavoro, 1991), ch. 4.

69 Steve Fraser, "'Landslayt' and 'Paesani': Ethnic Conflict and Cooperation in the Amalgamated Clothing Workers of America," in Dirk Hoerder (ed.), *"Struggle a Hard Battle": Essays on Working-Class Immigrants* (De Kalb: Northern Illinois University Press, 1986), pp. 280–303; Charles Anthony Zappia, "Unionism and the Italian American Workers, A History of the New York City 'Italian Locals' in the International Ladies Garment Workers Unions, 1900–1934," unpublished Ph.D. thesis, University of California Berkeley, 1994; Vezzosi, *Il socialismo indifferente*, pp. 178–82.

70 On passive Italian women in New York, Maxine Schwartz Seller, "The Uprising of the Twenty Thousand: Sex, Class, and Ethnicity in the Shirtwaist Makers' Strike of 1909," in Hoerder, *Struggle a Hard Battle*, pp. 254–78. On Italian women's activism, Colomba M. Furio, "The Cultural Background of the Italian Immigrant Woman and Its Impact on Her Unionization in the New York Garment Industry, 1880–1919," and Jean A. Scarpaci, "Angela Bambace and the International Ladies Garment Workers Union: the Search for an Elusive Activist," in George E. Pozzetta (eds), *Pane e Lavoro: The Italian American Working Class* (Toronto: Multicultural History Society of Ontario, 1980), pp. 353–9 and 99–118; Jennifer Guglielmo, "Defining Solidarity: Italian American Women's Workplace Organizing Strategies in the International Ladies' Garment Workers' Union during the Great Depression, New York City," in Donna Gabaccia and Franca Iacovetta (eds), *Foreign, Female and Fighting Back* (forthcoming). On Argentina, Carina Frid de Silberstein, "Becoming Visible: Italian Immigrant Women in the Garment and Textile Industries in Argentina, 1890–1930," in Gabaccia

and Iacovetta, *Foreign, Female and Fighting Back*; Matilde Alejandra, *La primera ley de trabajo femenino: "La Mujer Obrera" (1880–1910)* (Buenos Aires: Centro Ed. de America Latina, 1988).

71 Green, *Ready-to-Wear*, pp. 263–7.

72 Bruno Cartosio, "Gli emigrati italiani e l'Industrial Workers of the World," in Bezza, *Gli italiani fuori d'Italia*, pp. 359–97; Bruno Cartosio, "Sicilian Radicals in Two Worlds," in Debouzy, *In the Shadow of the Statue of Liberty*, pp. 127–38; Salvatore Salerno, *Red November, Black November; Culture and Community in the Industrial Workers of the World* (Albany: State University of New York Press, 1989), pp. 48–50; Michael Miller Topp, "The Transnationalism of the Italian-American Left: The Lawrence Strike of 1912 and the Italian Chamber of Labor of New York City," *Journal of American Ethnic History* 17, 1 (1997): 39–63; Michael Miller Topp, "The Lawrence Strike and the Defense Campaign for Ettor and Giovannitti: The Possibilities and Limitations of Italian American Syndicalism," in Donna Gabaccia and Fraser Ottanelli (eds), *For Us There Are No Frontiers* (forthcoming).

73 Ardis Cameron, *Radicals of the Worst Sort — Laboring Women in Lawrence, Massachusetts, 1860–1912* (Urbana: University of Illinois Press, 1993).

74 On the Austrian Social Democrats, see: Hans Mommsen, *Arbeiterbewegung und nazionale Frage* (Göttingen: Vandenhoeck und Ruprecht, 1979); Raimund Löw, "Der Zerfall der 'Kleinen Internationale': Nationalitätenkonflikte in der Arbeiterbewegung des alten Österreich (1889–1914)," *Materialien zur Arbeiterbewegung* 14 (Vienna: Europaverlag, 1984). On Trieste: Marina Cattaruzza, "L'emigrazione di forzalavoro verso Trieste dalla metà del XIX secolo alla prima guerra mondiale," *Movimento operaio e socialista* n.s. 1 (January–June 1978): 21–66; Marina Cattaruzza, *La formazione del proletariato urbano; Immigrati, operai di mestiere, donne a Trieste dalla metà del secolo XIX alla prima guerra mondiale* (Turin: Musolini, 1978). On Trieste's labor movement: Giuseppe Piemontese, *Il movimento operaio a Trieste, dalle origini all'avvento del fascismo* (Rome: Riuniti, 1974); Ennio Maserati, *Il sindacalismo autonomista triestino degli anni 1909–1914* (Trieste: Del Bianco, 1965); Ennio Maserati, *Il movimento operaio a Trieste dalle origini alla prima guerra mondiale* (Milan: Giuffrè, 1973); Ennio Maserati, *Gli anarchici a Trieste durante il dominio asburgico* (Milan: Giuffrè, 1977).

75 Pedroli, *Il socialismo nella Svizzera italiana*, p. 43; Anna Rosada, *Giacinto Menotti Serrati nell'emigrazione 1899–1911* (Rome: Riuniti, 1972), pp. 25–45; see also Nancy Eschelman, "Forging a Socialist Women's Movement: Angelica Balabanoff in Switzerland," in Caroli *et al.*, *Italian Immigrant Woman*, pp. 49–51.

76 Renato Monteleone, *Il movimento socialista nel Trentino 1894–1924* (Rome: Riuniti, 1971); see also S. Sutterlütti, "Italiener im Vorarlberg 1870–1914: Materielle Not und sozialer Widerstand," in *Im Prinzip: Hoffnung; Arbeiterbewegung im Vorarlberg, 1870–1946* (Bregenz: Fink's, 1984), pp. 133–57.

77 Michael M. Topp has made a similar argument for Italian radicals generally in "The Italian-American Left: Transnationalism and the Quest for Unity," in Paul Buhle and Dan Georgakas (eds), *The Immigrant Left in the United States* (Albany: State University of New York Press, 1996), pp. 119–47.

78 A summary of assassinations is in Arrigo Petacco, *L'anarchico che venne dall'America* (n.p., 1974) 1st edn 1969, p. 28.

79 On anti-anarchist campaigns, Richard Bach Jensen, "The International Anti-Anarchist Conference of 1898 and the Origins of Interpol," *Journal of Contemporary History* 16, 2 (1981): 323–47.

80 Good introductions to this world include Paul Avrich, *Sacco and Vanzetti: The Anarchist Background* (Princeton, N.J.: Princeton University Press, 1991); Rudolph J. Vecoli, "'Primo Maggio' in the United States," in Andrea Panaccione (ed.), *May Day Celebrations* (Venice: Marsilio, 1988), pp. 55–83; George W. Carey, "*La questione sociale*, An Anarchist Newspaper in Paterson, N.J. (1985–1908)," in Lydio Tomasi (ed.), *Italian Americans: New Perspectives in Italian Immigration and Ethnicity* (Staten Island: Center for Migration Studies, 1985), pp. 289–97.

81 Leonardo Bettini, *Bibliografia dell'anarchismo; Periodici e numeri unici anarchici in lingua italiana pubblicati all'estero (1872–1971)*, vol. I, tome 2 (Florence: CP, 1976); Franco Della Peruta, "Contributo alla bibliografia della stampa periodica operaia, anarchica e socialista, pubblicata all'estero in lingua italiana," *Movimento operaio* 2 (December–January 1949–1950): 103–4; Carlos Rama, "La stampa periodica italiana nell'America Latina," *Movimento operaio* 5 (September–October 1955): 802–5. For the U.S., see the Italian entry in Dirk Hoerder and Christiane Harzig (eds), *The Immigrant Labor Press in North America, 1840s–1970s* (Westport, Conn.: Greenwood Press, 1987), vol. 3; Augusta Molinari, "I giornali delle comunità anarchiche italo-americane," *Movimento operaio e socialista* n.s. 4 (January–June 1981): 117–30; Émile Témime, "Les journaux italiens à Marseille de 1870 à 1914," *Affari sociali internazionali* 5 (Fall–Winter 1977): 209–24.

82 Gabaccia, *Militants and Migrants*, pp. 138–40.

83 Donna Gabaccia, "Clase y cultura: Los migrantes italianos en los movimientos obreros en el mundo, 1876–1914," *Estudios migratorios Latinoamericanos* 7, 22 (1992): 433–5; David Montgomery, *The Fall of the House of Labor: The Workplace, the State, and American Labor Activism 1865–1925* (New York and Paris: Cambridge University Press and La Maison des Sciences de l'Homme, 1987), ch. 2; Eric J. Hobsbawm, *Primitive Rebels: Studies in Archaic Forms of Social Movement in the 19th and 20th Centuries* (Glencoe, Ill.: The Free Press, 1959), ch. 6. The direct quotation is from Bayer, "L'influenza dell'immigrazione italiana," p. 536.

84 Gary Mormino and George E. Pozzetta, *The Immigrant World of Ybor City: Italians and their Latin Neighbors in Tampa, 1885–1925* (Urbana: University of Illinois Press, 1987); Gary Mormino and George E. Pozzetta, "Concord and Discord: Italians and Ethnic Interactions in Tampa, Florida," in Lydio Tomasi (ed.), *Italian Americans: New Perspectives in Italian Immigration and Ethnicity* (New York: Center for Migration Studies, 1985), pp. 341–57; Durwood Long, "'La Resistencia': Tampa's Immigrant Labor Union," *Labor History* 6 (Fall 1965): 193–214; Anthony Pizzo, "The Italian Heritage in Tampa," in Robert F. Harney and J. Vincenza Scarpaci (eds), *Little Italies in North America* (Toronto: The Multicultural History Society of Ontario, 1981), pp. 123–40; George Pozzetta, "Italians and the General Strike of 1910," in Pozzetta,

Pane e Lavoro, pp. 429–36; George Pozzetta and Gary Mormino, "Immigrant Women in Tampa, the Italian Experience," *Florida Historical Quarterly* 61 (1983): 296–312.

85 Besides Ema Cibotti, "Mutualismo y politica en un estudio de caso: La Sociedad 'Unione e Benevolenza' en Buenos Aires entre 1858 y 1865," in Fernando J. Devoto and Gianfausto Rosoli (eds), *L'Italia nella società argentina* (Rome: Centro Studi Emigrazione, 1988), pp. 241–65, see Hilda Sabato and Ema Cibotti, "Hacer politica en Buenos Aires: Los italianos en las escena publica portena, 1860–1880," *Boletin del Instituto de Historia Argentina y Americana "Dr. E. Ravignani"* 3rd ser. 2 (1990): 20–3.

86 Samuel L. Baily, *Immigrants in the Lands of Promise: Italians in Buenos Aires and New York City, 1870–1914* (Ithaca, N.Y.: Cornell University Press, 1999), pp. 197–8; see also Luigi Favero, "Los Scalabrinianos y los emigrantes italianos en Sudamerica," *Estudios migratorios Latinoamericanos* 4, 12 (August 1989): 231–55; Gianfausto Rosoli, "Ordini, congregazioni religiose, chiesa e movimento operaio nell'emigrazione italiana in America Latina tra '800 e '900," in Blengino *et al.*, *La riscoperta delle Americhe*, pp. 444–71.

87 Romolo Gandolfo, "Los sociedades italianas de socorros mutuos de Buenos Aires: Cuestiones de etnicidad y de clase dentro de una comunidad de inmigrantes (1880–1920)," in Fernando J. Devoto and Eduardo J. Miguez (eds), *Asociacionismo, trabajo e identidad etnica; Los italianos en América Latina en una perspectiva comparada* (Buenos Aires: CEMLA-CSER-IEHS, 1992), pp. 311–32.

88 Mirta Lobato, "The *Patria degli italiani* and Social Conflict in Early Twentieth Century Argentina," in Gabaccia and Ottanelli, *For Us There are No Frontiers*.

89 Samuel L. Baily, "The Role of Two Newspapers and the Assimilation of Italians in Buenos Aires and São Paulo, 1893–1913," *International Migration Review* 12, 3 (1978): 321–40; Ema Cibotti, "Periodismo politico y politica periodistica; La construccion publica de una opinion italiano en el Buenos Aires finisecular," *Entrepasados, Revista de historia* 4, 7 (1994): 7–23.

90 Baily, *Immigrants in the Lands of Promise*, pp. 180–1.

91 The best brief overview is George E. Pozzetta, "The Mulberry District of New York City: The Years before World War One," in Harney and Scarpaci, *Little Italies in North America*, pp. 7–40; for the larger issues of class and community, see Robert F. Harney, "Ambiente e classi sociali nelle 'Little Italies' nordamericane," in Robert F. Harney, *Dalla frontiera alle Little Italies, Gli italiani in Canada 1800–1945* (Rome: Bonacci, 1984), pp. 197–213.

92 Jerre Mangione and Ben Morreale, *La Storia; Five Centuries of the Italian American Experience* (New York: HarperCollins, 1992), pp. 140–2; see also Green, "'Becoming American,'" pp. 162–4.

93 Pozzetta, "The Mulberry District," pp. 19–21.

94 Ibid., pp. 12–13.

95 On the importance of Columbus in the U.S., see Ferdinando Fasce, "L'immagine di Colombo negli Stati Uniti fra Otto e Novecento," in Ferdinando Fasce, *Tra due sponde; Lavoro, affari e cultura tra Italia e*

Stati Uniti nell'età della grande emigrazione (Genoa: Graphos, 1993), pp. 77–89.

96 The organization awaits its historian. For earlier works by activists, see Ernest Biagi, *The Purple Aster: A History of the Order Sons of Italy in America* (New York: Veritas Press, 1961); for the splits of the fascist years by an activist: Robert Ferrari, *Days Pleasant and Unpleasant in the Order Sons of Italy in America; The Problems of Races and Racial Societies in the United States. Assimilation or Isolation?* (New York: Mandy Press, 1926).

97 Gerald Meyer, *Vito Marcantonio, Radical Politican, 1902–1954* (Albany: State University of New York Press, 1989), ch. 7.

98 Mangione and Morreale, *La Storia*, p. 162.

99 The best introduction to the international dimensions of Vatican and clerical activism is in Gianfausto Rosoli, *Insieme oltre le frontiere* (Caltanissetta/Rome: Salvatore Sciascia Ed., 1996).

100 On the weakness of Catholicism among Italians, see Rudolph Vecoli, "Prelates and Peasants: Italian Immigrants and the Catholic Church," *Journal of Social History* 2 (1969): 217–68; Paul McBride, "The Italian-Americans and the Catholic Church: Old and New Perspectives, A Review Essay," *Italian Americana* 1, 2 (Spring 1975): 275.

101 On the importance of folk Catholicism see Robert A. Orsi, *The Madonna of 115th Street: Faith and Community in Italian Harlem, 1880–1950* (New Haven, Conn.: Yale University Press, 1985). For Catholic priests as leaders, Silvano M. Tomasi, *Piety and Power, The Role of the Italian Parishes in the New York Metropolitan Area, 1880–1930* (New York: Center for Migration Studies, 1975); Mary Elizabeth Brown, *From Italian Villages to Greenwich Village: Our Lady of Pompei, 1882–1992* (New York: Center for Migration Studies, 1992). On the priest's role in creating diaspora nationalism, see Peter R. D'Agostino, "The Scalabrini Fathers, the Italian Emigrant Church, and Ethnic Nationalism in America," *Religion and American Culture* 7, 1 (Winter 1997): 121–59.

102 Georges Haupt, "Il ruolo degli emigrati e dei rifugiati nella diffusione delle idee socialiste all'epoca della II Internazionale," in *Anna Kulischioff e l'età del riformismo* (Milan: Mondo Operaio-Avanti, 1978), pp. 60, 64.

6 NATION, EMPIRE, AND DIASPORA: FASCISM AND ITS OPPONENTS

1 Quoted in Maurizio Viroli, *For Love of Country; An Essay on Patriotism and Nationalism* (Oxford: Clarendon Press, 1997), p. 162.

2 Georges Haupt, *Socialism and the Great War, The Collapse of the Second International* (Oxford: Clarendon Press, 1972), p. 2.

3 G. Posta, "Perché non si deve emigrare," in *I quaderni delle corporazioni*, cited in Ornella Bianchi, "Fascismo ed emigrazione," in Vanni Blengino, Emilia Franzina, and Adolfo Pepe, *La riscoperta delle Americhe; Lavoratori e sindacato nell'emigrazione italiana in America Latina 1870–1970* (Milan: Teti Editore, 1993), p. 107n. On fascist emigration policy, see Philip V. Cannistraro and Gianfausto Rosoli, "Fascist Emigration Policy in the 1920s: An Interpretive Framework," *International Migration Review* 13 (1979): 673–92.

4 Benito Mussolini, *Opera omnia*, vol. 23, pp. 158–92, quoted in Richard Bosworth, *Italy and the Wider World* (New York: Routledge, 1996), p. 45.
5 Quoted in Philip V. Cannistraro, "Mussolini, Sacco–Vanzetti, and the Anarchists: The Transatlantic Context," *Journal of Modern History* 68, 1 (March 1996): 62.
6 Robert K. Murray, *Red Scare; A Study in National Hysteria, 1919–1920* (Minneapolis: University of Minnesota Press, 1955). On Italy's postwar radicalism and the rise of fascism, see Adrian Lyttelton, *The Seizure of Power: Fascism in Italy, 1919–1929* (London: Weidenfeld & Nicolson, 1987). On Argentina, John Raymond Hébert, *The Tragic Week in Buenos Aires: Background, Events, Aftermath* (Ann Arbor, Mich.: University Microfilms, 1974).
7 The best overview of nativist movements and immigration restriction in the U.S. remains John Higham, *Strangers in the Land: Patterns of American Nativism* (New York: Atheneum, 1965); see also Monte S. Finkelstein, "The Johnson Act, Mussolini and Fascist Emigration Policy, 1921–1930," *Journal of American Ethnic History* 8, 1 (Autumn 1988): 38–55. On Australia, see A.C. Palfreeman, *The Administration of the White Australia Policy* (Melbourne: Melbourne University Press, 1967). For Canada in this period, see Donald H. Avery, *Reluctant Host: Canada's Response to Immigrant Workers, 1896–1994* (Toronto: McClelland and Stewart, 1995).
8 Rudolf Schläpfer, *Die Ausländerfrage in der Schweiz vor dem Ersten Weltkrieg* (Zurich: Juris, 1969).
9 There is a large and growing literature on the origins of welfare states from comparative perspectives; see e.g. Peter Baldwin, *The Politics of Social Solidarity: Class Bases of the European Welfare States* (Cambridge: Cambridge University Press, 1990). A theoretical treatment of the relation of welfare state, migration, and integration of national labor movements is Gary Freeman, "Migration and the Political Economy of the Welfare State," *Annals of the American Academy of Political and Social Science* 485 (1986): 53–61. For the important case of Germany, see Knuth Dohse, *Ausländische Arbeiter und bürgerlicher Staat: Genese und Funktion von staatlicher Ausländerpolitik und Ausländerrecht* (Berlin: Express, 1985). Italy's government kept close watch over these developments, see Commissariato Generale dell'Emigrazione, *Il trattamento degli stranieri nei vari paesi; Condizione giuridica generale, cittadinanza e naturalizzazione; Regola di diritto internazionale privato; Assicurazioni sociali* (Rome: L'Universale, 1926).
10 *L'italiano all'estero, e la sua condizione giuridica, secondo le legislazioni straniere e gli accordi internazionali in vigore* (Rome: Direzione Generale degli Italiani all'Estero, 1934).
11 On interwar migrations to France, see Centre d'Étude et de Documentation sur l'Émigration Italienne, *L'Immigration italienne en France dans les années 20* (Paris: CEDEI, 1988); Pierre Milza, (ed.), *Les Italiens en France de 1914 à 1940* (Rome: Collection de l'École Française de Rome, 1986).
12 On Argentina, see Carl E. Solberg, *Immigration and Nationalism: Argentina and Chile, 1890–1914* (Austin: University of Texas Press, 1970); Emilio Gentile, "L'emigrazione italiana in Argentina nella politica di espansione del nazionalismo e del fascismo," *Storia contemporanea* 17 (June 1986): 371–9.

13 The representation of children among migrants — one rough measure of family relocation — increased modestly from 12 percent in 1916–25 to 19 percent in 1936–45; see Gianfausto Rosoli (ed.), *Un secolo di emigrazione italiana, 1876–1976* (Rome: Centro Studi Emigrazione, 1976).

14 See pp. 72–3 above (Chapter 3).

15 On U.S. migration patterns during these years, Thomas Archdeacon, *Becoming American: An Ethnic History* (New York: The Free Press, 1983), ch. 7. On Italian migration to the U.S., Claudia Belleri Damiani, "L'emigrazione italiana negli Stati Uniti durante il periodo fascista," in Renzo de Felice (ed.), *Cenni storici sulla emigrazione italiana nelle Americhe e in Australia* (Milan: Franco Angeli, 1979), pp. 105–23.

16 Rosoli, *Un secolo di emigrazione italiana.*

17 For background on Italy's policies, see Fernando Manzotti, *La polemica sull'emigrazione nell'Italia unita* (Milan: Dante Alighieri, 1969).

18 See Mario Benigni, "L'emigrazione italiana considerata da tre riviste cattoliche negli anni 1880–1914," in Franca Assante (ed.), *Il movimento migratorio italiano dall'unità nazionale ai nostri giorni* (Geneva: Librairie Droz, 1978), vol. 2, pp. 59–76; see also Edward C. Stabili, "The St. Raphael Society for the Protection of Italian Immigrants, 1887–1923," unpublished Ph.D. thesis, Notre Dame University, 1977.

19 Giovan Battista Sacchetti, "L'impegno sociale di Mons. G.B. Scalabrini e di Mons. G. Bonomelli nell'assistenza agli emigrati italiani: Caratteristiche e sviluppo storico," in Assante, *Movimento migratorio italiano*, vol. 2, pp. 85–105; Silvano M. Tomasi, "Scalabriniani e mondo cattolico di fronte all'emigrazione italiana (1880–1940)," in Bruno Bezza (ed.), *Gli italiani fuori d'Italia, Gli emigrati italiani nei movimenti operai dei paesi d'adozione 1880–1940* (Milan: Franco Angeli, 1983), pp. 145–62.

20 Luciano Trincia, *Emigrazione e diaspora; chiesa e lavoratori italiani in Svizzera e in Germania fino alla prima guerra mondiale* (Rome: Studium, 1997); Gianfausto Rosoli, "I movimenti di migrazione ed i cattolici," in Elio Guerriero and Annibale Zambarbieri (eds), *La chiesa e la società industriale (1878–1922)* (Milan: Paoline, 1990), pp. 497–526; Gianfausto Rosoli, "L'emigrazione italiana in Europa e l'Opera Bonomelli (1900–1914)," in Bezza, *Gli italiani fuori d'Italia*, pp. 163–202.

21 Italica Gens has not yet been the subject of intensive study but see Silvano Tomasi, "Fede e patria: the 'Italica Gens' in the United States and Canada, 1908–1936," *Studi emigrazione* 23 (1991): 319–41. For the older cultural organization, see Beatrice Pisa, *Nazione e politica nella Società "Dante Alighieri"* (Rome: Bonacci, 1995); Patrizia Salvetti, *Immagine nazionale ed emigrazione nella Società "Dante Alighieri"* (Rome: Bonacci, 1995).

22 See also Gerolamo Boccardo, *L'emigrazione e le colonie* (Florence: Le Monnier, 1871).

23 Giuseppe Are, *La scoperta dell'imperialismo; Il dibattito nella cultura italiana del primo Novecento* (Rome: Lavoro, 1985). For the view from Argentina, see Antonio Annino, "El debate sobre la emigracion y la expansion a la America Latina en los origines e la ideologia imperialista en Italia (1861–1911)," *Jahrbuch für Geschichte von Staat, Wirtschaft und Gesellschaft Lateinamerikas* 13 (1976): 192–5. See also Gigliola Dinucci, "Il modello della colonia libera nell'ideologia espansionista italiana: Dagli anni '80 alla fine del secolo," *Storia contemporanea* 10, 3 (1979): 427–79.

24 "Braccia che il paese protende lungi da sé su contrade estranee per trarle nell'orbita delle sue relazioni di lavoro e di scambio, devono esser come un allargamento della sua azione e della sua politica economica," quoted in Manzotti, *La polemica sull'emigrazione*, p. 66.
25 On Italy's early colonial ventures, see Tullio Scovazzi, *Assab, Massua, Uccialli, Adua: Gli strumenti giuridici del primo colonialismo italiano* (Turin: G. Giappichelli, 1996); Luigi Goglia, *Il colonialismo italiano da Adua all'impero* (Rome: Laterza, 1993).
26 Quoted in Claudio Segrè, *Fourth Shore: The Italian Colonization of Libya* (Chicago: University of Chicago Press, 1974), p. 18.
27 Ibid., p. 4.
28 Ronald S. Cunsoli, "Enrico Corradini and the Italian Version of Proletarian Nationalism," *Canadian Review of Studies in Nationalism* 12, 1 (1985): 47–63. For a general history, Alexander J. De Grand, *The Nationalist Association and the Rise of Fascism in Italy* (Lincoln: University of Nebraska Press, 1978).
29 Segrè, *Fourth Shore*; Angelo Del Boca, *Gli italiani in Africa orientale*, 3 vols. (Rome and Milan: Laterza and Mondadori, 1985–93).
30 Lucio Fabi, "La riforma della legge sulla cittadinanza italiana," *Gli italiani nel mondo* 21 (March 1965).
31 See note 8 above.
32 Luciano Tosi, "Italy and International Agreements of Emigration and Immigration," in *The World in My Hand, Italian Emigration in the World 1860/1960* (Rome: Centro Studi Emigrazione, 1997).
33 Commissariato Generale dell'Emigrazione, *Emigrazione e immigrazione; Considerazioni generali e documenti presentati alla conferenza internazionale dell'emigrazione e dell'immigrazione, Roma, maggio, 1924* (Rome: C. Colombo, 1924).
34 Cannistraro and Rosoli, "Fascist Emigration Policy," and Ornella Bianchi, "Fascismo ed emigrazione," in Blengino *et al.*, *La riscoperta delle Americhe*; see also "La nuova disciplina dell'emigrazione," *Bollettino dell'emigrazione* 7–8 (1927): 7–13, 7–10.
35 Vittorio Briani, "La legislazione italiana per la tutela dell'emigrazione," *Gli italiani nel mondo* 2, 15 (1946): 2–3; Tosi, "Italy and International Agreements on Emigration and Immigration," pp. 26–33. See also Joachim Lehmann, "Lavoratori italiani in Germania, 1938–39," *Ventesimo secolo* 2 (1991): 151–78.
36 Quoted in Zeffiro Ciuffoletti, *L'emigrazione nella storia d'Italia, 1868/1975, Storia e documenti* (Florence: Vallecchi, 1978), vol. 2, p. 177.
37 General introductions to the exile migrations include Charles F. Delzell, "The Italian Anti-Fascist Emigration, 1922–1943," *Journal of Central Europe Affairs* 12 (April 1952): 26–35; Francesca Taddei (ed.), *L'emigrazione socialista nella lotta contro il fascismo (1926–1939)* (Florence: Sansoni, 1982); Aldo Garosci, *Storia dei fuoriusciti* (Bari: Laterza, 1953); Santi Fedele, *I repubblicani in esilio: Nella lotta contro il fascismo (1926–1940)* (Florence: Le Monnier, 1989).
38 Philip V. Cannistraro and Gianfausto Rosoli, *Emigrazione, chiesa e fascismo; Lo scioglimento dell'Opera Bonomelli 1922–28* (Rome: Studium, 1979).
39 For Mussolini's agrarian policies, see: Domenico Preti, "La politica agraria del fascismo," *Studi storici* 14 (1973): 802–69; Jon S. Cohen, "Fascism and

Agriculture in Italy: Policies and Consequences," *Economic History Review* 32 (February 1979): 70–87. On internal migrations, see also p. 162 below.

40 Quoted in Ciuffoletti, *L'emigrazione nella storia d'Italia*, vol. 2, p. 156.

41 General studies include: Emilio Gentile, "La politica estera del partito fascista; Ideologia e organizzazione dei fasci italiani all'estero (1920–1930)," *Storia contemporanea* 26, 2 (December 1995): 897–956; Domenico Fabiano, "I fasci italiani all'estero," in Bezza, *Gli italiani fuori d'Italia*, pp. 221–36. On the role of the Vatican in aiding fascist propaganda abroad, see Gianfausto Rosoli, "Santa Sede e propaganda fascista all'estero tra i figli degli emigrati italiani," *Storia contemporanea* 17 (April 1986): 293–315.

42 Quoted in Angelo Del Boca, *Gli italiani in Libia: dal fascismo al Gheddafi* (Milan: Mondadori, 1997), p. 126.

43 Besides the works cited in note 34 above, see Gérard Crespo, *Les Italiens en Algérie, 1830–1960: Histoire et sociologie d'une migration* (Calvisson: J. Gandini, 1994).

44 Segrè, *Fourth Shore*, p. 4.

45 See biographies in Ugo E. Imperatori, *Dizionario di italiani all'estero (dal secolo XIII sino ad oggi)* (Genoa: L'Emigrante, 1956).

46 Quoted material from Philip V. Cannistraro, *Blackshirts in Little Italy: Italian Americans and Fascism, 1921–1929* (Lafayette, Ind.: Bordighera Press, 1999); Alan Cassels, "Fascism for Export: Italy and United States in the Twenties," *American Historical Review* 69, 3 (1964): 707–12.

47 Gaetano Salvemini, *Italian Fascist Activities in the United States*, ed. Philip V. Cannistraro (New York: Center for Migration Studies, 1977), pp. 244–5.

48 See biographies in Franco Andreucci and Tommaso Detti (eds), *Il movimento operaio italiano, Dizionario biografico 1853–1943* (Rome: Riuniti, 1975–8). On Sturzo and anti-fascist Catholic exiles, see Giuseppe Ignesti, "Momenti del popolarismo in esilio," in Pietro Scoppola and Francesco Traniello (eds), *I cattolici tra fascismo e democrazia* (Bologna: Il Mulino, 1975), pp. 75–183; Francesco Malgeri, "Il fuoruscitismo popolare," in *Storia del movimento cattolico in Italia*, vol. 4: *I cattolici dal fascismo alla resistenza* (Rome: Il Poligono, 1981); Franco Rizzi, "Sturzo in esilio; Popolari e forze antifasciste dal 1924 al 1940," in *Luigi Sturzo nella storia d'Italia; Atti del Convegno internazionale di studi promosso dall'Assemblea Regionale Siciliana* (Rome: Storia e Letterature, 1973), pp. 499–567.

49 Philip Cannistraro, "Luigi Antonini and the Italian Anti-Fascist Movement in the United States, 1940–1943," *Journal of American Ethnic History* 5 (Fall 1985): 21–40.

50 Fraser Ottanelli, "Italian-American Antifascism and the Remaking of the U.S. Working Class," in Donna Gabaccia and Fraser Ottanelli, *For Us There Are No Frontiers* (forthcoming).

51 Salvemini, *Italian Fascist Activities*, pt. 2.

52 Madeline J. Goodman, "The Evolution of Ethnicity: Fascism and Anti-Fascism in the Italian-American Community," unpublished Ph.D. thesis, Carnegie Mellon University, 1993; Vincent M. Lombardi, "Italian American Workers and the Response to Fascism," in George E. Pozzetta (ed.), *Pane e Lavoro: The Italian American Working Class* (Toronto: Multicultural History Society of Ontario, 1980), pp. 141–58; Philip

Cannistraro, "Generoso Pope and the Rise of Italian American Politics, 1925–1935," in Lydio Tomasi (ed.), *Italian Americans: New Perspectives in Italian Immigration and Ethnicity* (Staten Island: Center for Migration Studies, 1985), pp. 264–88.

53 Philip V. Cannistraro, "Mussolini, Sacco–Vanzetti, and the Anarchists: The Transatlantic Context," *Journal of Modern History* 68, 1 (March 1996): 31–62. See also Nunzio Pernicone, "Carlo Tresca and the Sacco–Vanzetti Case," *Journal of American History* 66 (December 1979): 535–47; Paul Avrich, *Sacco and Vanzetti: The Anarchist Background* (Princeton, N.J.: Princeton University Press, 1992).

54 John P. Diggins, *Mussolini and Fascism; The View from America* (Princeton, N.J.: Princeton University Press, 1972).

55 Rudolph J. Vecoli, "The Interwar Years," in Philip V. Cannistraro (ed.), *The Lost World of Italian-American Radicalism* (forthcoming); for a similar argument for Canada, see Roberto Perin, "Making Good Fascists and Good Canadians: Consular Propaganda and the Italian Community in Montreal in the 1930s," in Gerald Gold (ed.), *Minorities and Mother Country Imagery* (St. Johns: Institute of Social and Economic Research, Memorial University of Newfoundland, 1984).

56 Peter R. D'Agostino, "The Scalabrini Fathers, the Italian Emigrant Church and Ethnic Nationalism in America," *Religion and American Culture* 7, 1 (Winter 1997): 149. See also Peter R. D'Agostino, "The Triad of Roman Authority: Fascism, the Vatican, and the Italian Religious Clergy in the Italian Emigrant Church," *Journal of American Ethnic History* 17, 3 (Spring 1998): 3–37.

57 Nadia Venturini, *Neri e italiani ad Harlem: Gli anni trenta e la guerra d'Etiopia* (Rome: Lavoro, 1990); also useful for these years in Harlem is Ronald Bayor, *Neighbors in Conflict: The Irish, Germans, Jews and Italians of New York City, 1929–1941* (Baltimore, Md.: Johns Hopkins University Press, 1978).

58 William R. Scott, "Black Nationalism and the Italo–Ethiopian Conflict," *Journal of Negro History* 63 (April 1978): 118–34; for background, Cheryl L. Greenberg, *Or Does it Explode?: Black Harlem in the Great Depression* (New York: Oxford University Press, 1991).

59 Stephen C. Fox, *The Unknown Internment: An Oral History of the Relocation of Italian Americans during World War II* (Boston, Mass.: Twayne Publishers, 1990). For Canada, see Franca Iacovetta, Roberto Perin, and Angelo Principe, *Enemies Within: Canadian and International Studies on Italian and Other Wartime Internees* (Toronto: University of Toronto Press, forthcoming); for Australia, Richard Bosworth and Romano Ugolino (eds), *War, Internment and Mass Migration: The Italo–Australian Experience, 1940–1990* (Rome: Gruppo ed. Internazionale, 1992).

60 Nunzio Pernicone, "Carlo Tresca: Life and Death of a Revolutionary," in Richard N. Juliani and Philip V. Cannistraro (eds), *Italian Americans: The Search for a Useable Past* (New York: Center for Migration Studies, 1989).

61 A good general introduction in English remains Charles F. Delzell, *Mussolini's Enemies: The Italian Anti-Fascist Resistance* (Princeton, N.J.: Princeton University Press, 1961); on Belgium, see Anne Morelli, *Fascismo e antifascismo nell'emigrazione italiana in Belgio (1922–1940)* (Rome: Bonacci, 1987).

62 Pierre Milza, *Le Fascisme italien et la presse française: 1920–1940* (Brussels: Complexe, 1987).

63 Pierre Guillen, "Le rôle politique de l'immigration italienne en France dans l'entre-deux guerres," in Milza, *Les Italiens en France*, pp. 337–41; Eric Vial, "Notes sur l'exil et l'intégration des italiens dans la société française pendant le fascisme," in Maurizio Degl'Innocenti (ed.), *L'esilio nella storia del movimento operaio e l'emigrazione economica* (Manduria: Piero Lacaito, 1992), pp. 171–84; Pierre Guillen, "L'antifascisme, facteur d'intégration des Italiens en France dans l'entre-deux-guerres," in Taddei, *L'emigrazione socialista*, pp. 209–20.

64 Pierre Milza, *Voyage en Ritalie* (Paris: Plon, 1993), pp. 352–4, 261–4; Morelli, *Fascismo e antifascismo*, pp. 60–78.

65 Milza, *Voyage*, pp. 235–51; Guillen, "Le rôle," p. 325; Morelli, *Fascismo e antifascismo*, pp. 82–90.

66 Besides Ottanelli, "Italian-American Antifascism," see Nunzio Pernicone, "Italian Antifascists in the Spanish Civil War: The Battle of Guadalajara," *La parola del popolo* (November–December 1978): 132–40.

67 Carlo Rosselli, *Oggi in Spagna, domani in Italia* (Paris: Giustizia e Libertà, 1938).

68 Guillen, "L'antifascisme," pp. 218–20; Guillen, "Le rôle," pp. 336–7; Ralph Schor, "Les italiens dans les Alpes-Maritimes," in Milza, *Les Italiens en France*, pp. 595–8; Rudy Damiani, "Les italiens dans le nord et le Pas-de-Calais," in Milza, *Les Italiens en France*, p. 657.

69 Paul Ginsborg, *A History of Contemporary Italy; Society and Politics, 1943–1988* (London: Penguin, 1990), p. 196.

70 On fascism and anti-fascism in Canada, see John Zucchi, *Italians in Toronto: Development of a National Identity 1875–1935* (Montreal and Kingston: McGill-Queen's University Press, 1988), ch. 7; Robert F. Harney, "La 'Little Italy' di Toronto 1919–1945," in Robert F. Harney, *Dalla frontiera alle Little Italies: Gli italiani in Canada 1800–1945* (Rome: Bonacci, 1984); Luigi Bruti Liberati, *Il Canada, l'Italia e il fascismo, 1919–1945* (Rome: Bonacci, 1984). On Australia, Gianfranco Cresciani, *Fascism, Anti-fascism, and Italians in Australia, 1922–1945* (Canberra: Australian National University Press, 1980); Gianfranco Cresciani, "I socialisti italiani in Australia," in Taddei, *L'emigrazione socialista*, pp. 293–303; Gianfranco Cresciani, "The Proletarian Migrants: Fascism and Italian Anarchists in Australia," *Australian Quarterly* 51, 1 (1979): 4–19.

71 Angelo Trento, "Le associazioni italiane a São Paulo, 1878–1960," in Fernando J. Devoto and Eduardo J. Miguez (eds), *Asociacionismo, trabajo e identidad etnica; Los Américanos en América Latina en una perspectiva comparada* (Buenos Aires: CEMLA-CSER-IEHS, 1992), pp. 31–59; and Angelo Trento, "Il Brasile, gli immigrati e il fenomeno fascista," in Vanni Blengino, Emilia Franzina, and Adolfo Pepe, *La riscoperta delle Americhe, Lavoratori e sindacato nell'emigrazione italiana in America Latina 1870–1970* (Milan: Teti Editore, 1993), pp. 250–64; Chiara Vangelista, "Una revista italiana e a immigraçao no Brasil durante os primeiro anos do Fascismo: Le vie d'Italia e dell'America Latina, 1924–32," *Veritas* 40 (1995): 243–54; João Fabio Bertonha, "La base sociale dell'antifascismo a São Paulo; un'analisi 1923–1930," in Blengino *et al.*, *La riscoperta delle Americhe*, pp. 390–9.

72 Maria de Lujén Leiva, "Il movimento antifascista italiano in argentina (1922–1945)," in Bezza, *Gli italiani fuori d'Italia*, pp. 549–82; Pietro Rinaldo Fanesi, "L'esilio antifascista e la comunità italiana in Argentina," in Blengino *et al.*, *La riscoperta delle Americhe*, pp. 115–31; María Ostuni, "Operai e antifascismo a Buenos Aires: La società 'Liber Piemont,'" in Devoto and Miguez, *Asociacionismo, trabajo e identidad etnica*, pp. 303–10; Pietro Rinaldo Fanesi, *Verso l'altra Italia; Albano Corneli e l'esilio antifascista in Argentina* (Milan: Franco Angeli, 1991); Emilio Gentile, "Emigración e italianidad en Argentina en los mitos de potencia del nacionalismo y del fascismo (1900–1930)," *Estudios migratorios Latinoamericanos* 1, 2 (1986): 143–80; Irene Guerrini and Marco Pluviano, "L'organizzazione del tempo libero nelle comunità italiane in America Latina: l'Opera Nazionale Dopolavoro," in Blengino *et al.*, *La riscoperta delle Americhe*, pp. 378–89.

7 POSTWAR ITALY: FROM SENDING TO RECEIVING NATION

1 A New York City immigrant, cited in Thomas Kessner and Betty Boyd Caroli, *Today's Immigrants: Their Stories* (New York: Oxford University Press, 1982), p. 214.
2 Useful general histories, with good attention to politics, include Paul Ginsborg, *A History of Contemporary Italy; Society and Politics, 1954–1988* (London: Penguin, 1990); Silvio Lanaro, *Storia dell'Italia repubblicana* (Venice: Marsilio, 1992); Aurelio Lepre, *Storia della prima repubblica* (Bologna: Il Mulino, 1993).
3 Quoted in Richard Bosworth, *Italy and the Wider World* (New York: Routledge, 1996), p. 123.
4 Ugo Ascoli, *Welfare State all'italiana*, 2nd edn (Rome: Laterza, 1984).
5 There are few general studies of the postwar migrations, but see Anna Maria Birindelli, "Stabilità e mutamenti della dinamica migratoria italiana all'estero negli ultimi decenni," in Fernando J. Devoto and Gianfausto Rosoli (eds), *L'Italia nella società argentina* (Rome: Centro Studi Emigrazione, 1988), pp. 102–23; G. Lucrezio Monticelli, "La dinamica dell'emigrazione italiana nel dopoguerra," *Studi emigrazione* 1, 3 (1965): 3–15.
6 Sidney Pollard, *The International Economy since 1945* (London: Routledge, 1997), Table 1.1.
7 Ginsborg, *A History of Contemporary Italy*, pp. 210–12.
8 Sydney Tarrow, *Peasant Communism in Southern Italy* (New Haven, Conn.: Yale University Press, 1967); Paolo Cinanni, *Lotte per la terra nel Mezzogiorno, 1943–1953: Terre pubbliche e trasformazione agraria* (Venice: Marsilio, 1979); Piero Bevilacqua, *Le campagne del Mezzogiorno tra fascismo e dopoguerra: Il caso della Calabria* (Turin: G. Einaudi, 1980).
9 James E. Miller, *The United States and Italy, 1940–1950* (Chapel Hill: University of North Carolina Press, 1986), pp. 213–74; James E. Miller, "Taking Off the Gloves: The United States and the Italian Elections of 1948," *Diplomatic History* 7, 1 (1983): 34–55.
10 John L. Harper, *America and the Reconstruction of Italy* (Cambridge: Cambridge University Press, 1986).
11 For an overview of the postwar South in English, see Alan B. Mountjoy, *The Mezzogiorno* (London: Oxford University Press, 1982). For a sense of

southern life in the postwar period, see Judith Chubb, *Patronage, Power and Poverty in Southern Italy; A Tale of Two Cities* (Cambridge: Cambridge University Press, 1982).

12 Russell King, *Land Reform: The Italian Experience* (London: Butterworths, 1973); Manlio Rossi Doria, *Dieci anni di politica agraria nel Mezzogiorno* (Bari: Laterza, 1958).

13 Gianfausto Rosoli (ed.), *Un secolo di emigrazione italiana, 1876–1976* (Rome: Centro Studi Emigrazione, 1978).

14 From an uneven historiography, see, on the U.S., David M. Reimers, *Still the Golden Door: The Third World Comes to America* (New York: Columbia University Press, 1985), ch. 1. For Canada, Freda Hawkins, *Canada and Immigration: Public Policy and Public Concern*, 2nd edn (Toronto: Institute of Public Administration of Canada, 1988); see also Freda Hawkins, *Critical Years in Immigration: Canada and Australia Compared* (Montreal: McGill-Queen's University Press, 1991). Little has been written about Argentina's postwar policies or migrations but see Silvia Lepore, "Migración italiana y politica migratoria argentina (1976–1989)," *Estudios migratorios Latinoamericanos* 11 (1989): 159–97, and Silvia Lepore, "Economic Profile of Italian Argentinians in the 1980s," in Lydio Tomasi, Piero Gastaldo, and Thomas Row (eds), *The Columbus People: Perspectives in Italian Emigration to the Americas and Australia* (New York and Turin: Center for Migration Studies and Fondazione Giovanni Agnelli, 1994), pp. 125–51. See also Susan Berglund, "Italian Immigration in Venezuela: A Story Still Untold," in Tomasi *et al.*, *The Columbus People*, pp. 173–209.

15 Luciano Tosi, "Italy and International Agreements on Emigration and Immigration," in *The World in My Hand, Italian Emigration in the World 1860/1960* (Rome: Centro Studi Emigrazione, 1997).

16 M. Gianturco, quoted in Bosworth, *Italy and the Wider World*, p. 127.

17 Paola Salvatori, "Politica sindacale per l'emigrazione nel secondo dopoguerra," in Vanni Blengino, Emilia Franzina, and Adolfo Pepe (eds), *La riscoperta delle Americhe; Lavoratori e sindacato nell'emigrazione italiana in America Latina 1870–1970* (Milan: Teti Editore, 1993), pp. 132–46.

18 Besides Tosi, "Italy and International Agreements," see Lorenzo Bertucelli, "Politica emigratoria e politica estera: Il ruolo del Sindacato," in Blengino *et al.*, *La riscoperta delle Americhe*, pp. 147–67.

19 Tosi, "Italy and International Agreements." Helpful for understanding the postwar context is Martin Baldwin-Edwards and Martin A. Schain (eds), *The Politics of Immigration in Western Europe* (London: Frank Cass, 1994).

20 Ministro degli Affari Esteri, *L'italiano nel mondo e la sua condizione giuridica secondo le legislazioni straniere e gli accordi internazionali* (Rome: Istituto Poligrafico dello Stato, 1954).

21 Jytte Klausen and Louise A. Tilly, "European Integration in a Social and Historical Perspective," in Jytte Klausen and Louise A. Tilly (eds), *European Integration in Social and Historical Perspective, 1850 to the Present* (Lanham: Rowman & Littlefield, 1997), pp. 6–8.

22 Rosoli, *Un secolo di emigrazione italiana*, Table 7.

23 Nora Federici, "Aspetti quasi ignorati dell'assistenza nell'emigrazione," *Italiani nel mondo* 11, 18 (1955): 16–17.

24 On the interaction of paese and state auspices for migration to Australia, see J.S. MacDonald, "Italian Migration to Australia: Manifest Functions of

Bureaucracy Versus Latent Functions of Informal Networks," *Journal of Social History* 3 (1970): 248–76.

25 Rosoli, *Un secolo dell'emigrazione italiana.*

26 The story of the Donatos is from Franca Iacovetta, *Such Hardworking People: Italian Immigrants in Postwar Toronto* (Montreal: McGill-Queen's University Press, 1992), p. 3; see also Franca Iacovetta, "'Primitive Villagers and Uneducated Girls': Canada Recruits Domestics from Italy, 1951–1952," *Canadian Woman Studies* 7 (1987); Franca Iacovetta, "Ordering in Bulk: Canada's Postwar Immigration Policy and the Recruitment of Contract Workers from Italy," *Journal of American Ethnic History* 11, 1 (1991): 50–80. See also Franc Sturino, "Contours of Postwar Italian Immigration to Toronto," *Polyphony* 6, 1 (Spring/Summer 1984): 127–30.

27 Iacovetta, *Such Hardworking People*, Tables 12 and 14.

28 Ibid., chs. 6 and 7.

29 For the origins of European union, see Klausen and Tilly, "European Integration in a Social and Historical Perspective"; Alan S. Milward, *The Reconstruction of Western Europe, 1945–1951* (Berkeley: University of California Press, 1984); Alan S. Milward, *The European Rescue of the Nation-State* (Berkeley: University of California Press, 1992).

30 "L'accordo per l'emigrazione con la repubblica di Bonn," *Italiani nel mondo* 11, 14 (1955). For an historian's perspective on guest workers, see Leslie Page Moch, "Foreign Workers in Western Europe: The 'Cheaper Hands' in Historical Perspective," in Klausen and Tilly, *European Integration*, pp. 103–16, and Ulrich Herbert, *A History of Foreign Labor in Germany, 1880–1980: Seasonal Workers, Forced Laborers, Guest Workers* (Ann Arbor: University of Michigan Press, 1990).

31 Andrea Leonardi, Alberto Cova, and Pasquale Galea, *Il Novecento economico italiano: Dalla grande guerra al "miracolo economico": 1914–1962* (Bologna: Monduzzi, 1997). For the social consequences, see Guido Crainz, *Storia del miracolo italiano: Cultura, identità, trasformazioni fra anni cinquanta e sessanta* (Rome: Donzelli, 1996).

32 Pasquale Saraceno, *L'Italia verso la piena occupazione*, quoted in Ginsborg, *A History of Contemporary Italy*, p. 229.

33 See ibid., chs. 10–11, and "Many Mountains Still to Climb," *The Economist*, November 8, 1997.

34 For the comparison to Germany, see Klaus Bade, *Vom Auswanderungsland zum Einwanderungsland? Deutschland, 1880–1980* (Berlin: Colloquium, 1983). Like the postwar migrations generally, Italy's transition has not yet found its historian, but see Emilio Reyneri, *La catena migratoria; Il ruolo dell'emigrazione nel mercato del lavoro di arrivo e di esodo* (Bologna: Il Mulino, 1979); Federico Romero, *L'emigrazione e integrazione europea, 1945–1973* (Rome: Lavoro, 1991); in English see Ginsborg, *A History of Contemporary Italy*, ch. 7.

35 Commissariato per le Migrazioni e la Colonizzazione, *Le migrazioni interne in Italia* (Rome: Poligrafico Italiano, 1931–35); Ercole Sori, "Emigrazione all'estero e migrazioni interne in Italia fra le due guerre," *Quaderni storici* 29/30 (1975): 579–607; Anna Treves, "The Anti-urban Policy of Fascism and a Century of Resistance to Industrial Urbanization in Italy," *International Journal of Urban and Regional Research* 4 (1980): 470–84; Anna Treves, *Le migrazioni interne nell'Italia fascista; Politica e realtà demografica*

(Turin: Einaudi, 1976); Corrado Grassi, *Le migrazioni interne italiane nel secolo unitario; Cause e conseguenze* (Turin: G. Giappichelli, 1967).

36 Quoted in Zeffiro Ciuffoletti, *L'emigrazione nella storia d'Italia, 1868/1975, Storia e documenti* (Florence: Vallecchi, 1978), vol. 2, p. 147.

37 Federico Romero, "Da emigrati in America a gastarbeiter: L'emergere dell'interdipendenza dei mercati del lavoro negli anni sessanta," in Blengino *et al.*, *La riscoperta delle Americhe*, pp. 182–97.

38 For Italian migrations to Germany, see Barbara von Breitenbach, *Italiener und Spanier als Arbeitnehmer in der Bundesrepublik Deutschland: Eine vergleichende Untersuchung zur europäischen Arbeitsmigration* (Munich/Mainz: Kaiser, Grunewald, 1982); Pietro Coletto, *I lavoratori italiani nella repubblica federale di Germania* (Milan: Univ. Cattolica del Sacro Cuore, Fac. di Scienze Politiche, 1965–6); Luigi Cajani and Brunello Mantelli, "In Deutschland arbeiten: Die Italiener — von der 'Achse' bis zur Europäischen Gemeinschaft," *Archiv für Sozialgeschichte* 32 (1992): 231–46.

39 On the treaty, see Leonida Felletti, "L'accordo italo-svizzero e una china pericolosa," *Italiani nel mondo* 21, 1 (1965); Leonida Felletti, "Dopo una illustrazione di Storchi: La camera approva l'accordo di emigrazione italo-svizzero," *Italiani nel mondo* 21, 2 (1965). On Italian workers in Switzerland, see Rudolf Braun, *Sozio-kulturelle Probleme der Eingliederung italienischer Arbeitskräfte in der Schweiz* (Erlenback-Zurich: Eugen Rentsch, 1970); Lucio Boscardin, *Die italienische Einwanderung in die Schweiz mit besonderer Berücksichtigung der Jahre 1946–1959* (Zurich: Polygraphischer, 1962); Delia Castelnuovo Frigessi, *Elvezia, il tuo governo: Operai italiani emigrati in Svizzera* (Turin: G. Einaudi, 1977).

40 The best source for the internal migrants and their characteristics is Ugo Ascoli, *Movimenti migratori in Italia* (Bologna: Il Mulino, 1979).

41 Rosoli, *Un secolo di emigrazione italiana.*

42 Ibid.

43 *Italiani nel mondo* 1, 1 (1945).

44 Most studies focus on brain-drain migrations from particular Third World countries, or on the problems they pose for economic development. But such migrations are not limited to those from poorer countries. For an overview, see F.J. van Hoek, *The Migration of High Level Manpower from Developing to Developed Countries* (The Hague: Mouton, 1970).

45 My analysis of 576 elite migrants from Ugo E. Imperatori, *Dizionario di italiani all'estero (dal secolo XIII sino ad oggi)* (Genoa: L'Emigrante, 1956).

46 See biographies in ibid.

47 *Italiani nel mondo* 2, 12 (1946).

48 See reports on the first world congresses, *Italiani nel mondo*, 31, 5–6 (March 1975); 7–8 (April 1975). See also "CCIE senza pace," *Italiani nel mondo* 31, 19–20 (October 1975). Quoted material from "La proposta di legge delle associazioni sui comitati consolari," *Italiani nel mondo* 36, 1–2 (January 1980).

49 Kessner and Caroli, *Today's Immigrants*, pp. 230–1, 219, 221.

50 The best general source (and the source of all the statistics cited here) remains Stephen Castles and Godula Kosack, *Immigrant Workers and Class Structure in Western Europe*, 2nd edn (New York: Oxford University Press, 1985), Tables III:6; III:8; III:11.

51 In Germany particularly, a great deal of attention focused on the criminality of the foreign workers. See Hans-Bernd Grüber, *Kriminalität der Gastarbeiter; Zusammenhang zwischen kulturellem Konflikt und Kriminalität* (Hamburg: Dissertation, 1969).
52 For the particularly rigid labor policies of Switzerland, see Hans-Joachim Hoffman-Nowotny, "Switzerland: A Non-Immigration Immigration Country," in Robin Cohen (ed.), *The Cambridge Survey of World Migration* (Cambridge: Cambridge University Press, 1995), pp. 302–7; Franco Pittau, *Emigrazione italiana in Svizzera: Problemi del lavoro e della sicurezza sociale* (Milan: Franco Angeli, 1984); Claudio Calvaruso, *Sottoproletariato in Svizzera: 152,000 lavoratori stagionali, Perché?* (Rome: Coines, 1971).
53 "Lavoratori stranieri nella realtà svizzera," *Italiani nel mondo* 21 (1975).
54 Numbers are from Rosemarie Rogers (ed.), *Guests Come to Stay: The Effects of European Labor Migration on Sending and Receiving Countries* (Boulder, Colo.: Westview Press, 1985), Table 1–1.
55 See "Postscript to the Second Edition, 1983," in Castles and Kosack, *Immigrant Workers*; Thomas Faist, "Migration in Contemporary Europe: European Integration, Economic Liberalization, and Protection," in Klausen and Tilly, *European Integration*, pp. 223–48. On the concept of denizen, see Robin Cohen, *Frontiers of Identity; The British and the Others* (London: Longman, 1994), p. 189.
56 For early use of the term immigrant for internal migrants, see: Centro di Ricerche Industriali e Sociali di Torino (CRIS), *Immigrazione e industria* (Milan: Comunità, 1962); Celestino Canteri, *Immigrati a Torino* (Milan: Avanti!, 1962).
57 Besides Ascoli, *Movimenti migratori*, ch. 3, general studies include Adriano Baglivo and Giovanni Pellicciari, *Sud amaro: Esodo come sopravvivenza* (Milan: Sapere, 1970); A. Corsi, "L'esodo agricolo dagli anni '50 agli anni '70 in Italia e nel Mezzogiorno," *Rassegna economica* 41, 3 (1977): 721–53.
58 Mario Walter Battacchi, *Meridionali e settentrionali nella struttura del pregiudizio etnico in Italia* (Bologna: Il Mulino, 1972).
59 Ginsborg, *A History of Contemporary Italy*, pp. 223–7.
60 Giovanni Pellicciari (ed.), *L'immigrazione nel triangolo industriale* (Milan: Franco Angeli, 1970), ch. 6. See also Francesco Compagna, *I terroni in città* (Bari: G. Laterza, 1959).
61 Ginsborg, *A History of Contemporary Italy*, pp. 250–3.
62 Pellicciari, *L'immigrazione nel triangolo industriale*, p. 342.
63 The numbers are from Demetrios G. Papademetriou, *Coming Together or Pulling Apart? The European Union's Struggle with Immigration and Asylum* (Washington, D.C.: Carnegie Endowment for International Peace, 1996), Tables 2–3. General studies include: Nora Federici (ed.), "L'immigrazione straniera in Italia," special issue, *Studi emigrazione* 71 (1983); André Jacques, *Lo straniero in mezzo da noi: Gli sradicati nel mondo d'oggi: La situazione in Italia* (Turin: Claudiana, 1987).
64 Carla Collicelli, "Immigration and Cultural Anxiety in Italy," *Affari sociali internazionali* 23, 2 (1995); Marco Jacquemet, "The Discourse on Migration and Racism in Contemporary Italy," in Salvatore Sechi (ed.), *Deconstructing Italy: Italy in the Nineties* (Berkeley, Calif.: International and Area Studies, 1995).

65 See, e.g., Francesca Villa (ed.), *Immigrati extracommunitari a Milano e in Lombardia: Atti del corso di aggiornamento della Fondazione Verga* (Milan: Vita e Pensiero, 1990); Ode Barsotti (ed.), *La presenza straniera in Italia: Il caso della Toscana* (Milan: F. Angeli, 1988); Anna Maria Birindelli, *La presenza straniera in Italia: Il caso dell'area romana* (Milan: Franco Angeli, 1993).
66 Salah Methnani, *Immigrato* (1990), quoted in Pasquale Verdicchio, *Bound by Distance: Rethinking Nationalism through the Italian Diaspora* (Madison, N.J.: Fairleigh Dickinson University Press, 1997), p. 157.
67 ECAP–CGIL, EMIM (ed.), *Immigrazione straniera*, cited in Francesco Carchedi and Giovanni Battista Ranuzzi, "Tra collocazione nel mercato del lavoro secondario ed esclusione dal sistema della cittadinanza," in Nino Sergi (ed.), *L'immigrazione straniera in Italia* (Rome: Lavoro, 1987), p. 129. A burst of publications examining the legal status of foreigners in Italy began appearing; see Nicolo Amato, *Casa o fortezza?: L'Italia, l'Europa del 1992 e l'immigrazione: Quali scelte politiche?* (Turin: Claudiana, 1989); Giustino D'Orazio, *Lo straniero nella costituzione italiana: Asilo, condizione giuridica, estradizione* (Padova: CEDAM, 1992).
68 Demetrios G. Papademetriou and Kimberly A. Hamilton, *Converging Paths to Restriction: French, Italian, and British Responses to Immigration* (Washington, D.C.: Carnegie Endowment for International Peace, 1996), pp. 41–2.
69 Ibid., ch. 3.
70 The use of the phrase in this way originates in British studies. See *The Empire Strikes Back: Race and Racism in 70s Britain* (London: Centre for Contemporary Cultural Studies, University of Birmingham, 1982).
71 Vanessa Maher, "Immigration and Social Identities," in Robert Lumley and Jonathan Morris (eds), *The New History of the Italian South; The Mezzogiorno Revisited* (Exeter: University of Exeter Press, 1997), pp. 160–4.
72 "Per gli emigrati," and "Dalle regioni," *Italiani nel mondo* 36, 7–8 (May 1980).
73 Amalia Signorelli, "Movimenti di popolazione e trasformazioni culturali," in Francesco Barbagallo (ed.), *Storia dell'Italia repubblicana*, vol. 2: *La trasformazione dell'Italia: Sviluppo e squilibri, part. 1, Politica, economia, società* (Turin: Giulio Einaudi, 1995), p. 641.
74 Susi Fantino, "Emigrazione di ritorno: Due identità a confronto: Gli argentini delle Langhe," in Blengino *et al.*, *La riscoperta delle Americhe*, pp. 644–64.
75 Franco Pittau and Nino Sergi (eds), *Emigrazione e immigrazione: Nuove solidarietà* (Rome: Lavoro, 1989).
76 Verdicchio, *Bound by Distance*, p. 153.
77 Mark J. Miller, "Political Participation and Representation of Noncitizens," in William Rogers Brubaker (ed.), *Immigration and the Politics of Citizenship in Europe and North America* (Lanham: University Press of America, 1989), p. 135.

8 CIVILTÀ ITALIANA AND THE MAKING OF MULTI-ETHNIC NATIONS

1 Giuseppe Pitrè, *Proverbi siciliani* (Palermo: "Il Vespro," 1978), vol. 3, pp. 118–19.

2 Increasingly, scholars observe that globalization also turns people back toward the particular, the regional, and the local dimensions of their lives; see Wang Gungwu, "Migration History: Some Patterns Revisited," Wang Gungwu (ed.), *Global History and Migrations* (Boulder, Colo.: Westview Press, 1997), p. 8.

3 I borrow this typology of diasporas — trade, labor, cultural — from Robin Cohen, *Global Diasporas, An Introduction* (Seattle: University of Washington Press, 1997), p. ix. What I call "exile diasporas" resemble Cohen's "victim diasporas" but had much shorter lives than the ones Cohen describes (Jews, Armenians, Africans).

4 Pilsudski is quoted in E.J. Hobsbawm, *Nations and Nationalism since 1780: Programme, Myth, Reality* (Cambridge: Cambridge University Press, 1990), pp. 44–5.

5 Italy's new governments since 1992 have remained reluctant to extend voting rights to citizens living abroad.

6 The typology of national identities is from Stephen Castles and Mark J. Miller, *The Age of Migration; International Population Movements in the Modern World* (New York: The Guilford Press, 1993), ch. 8; here I offer an only slightly modified version of the interpretation developed in Donna Gabaccia and Fraser Ottanelli, "Diaspora or International Proletariat?," *Diaspora* 6, 1 (1997): 61–834.

7 Gianfausto Rosoli, "Le popolazioni di origine italiana oltreoceano," *Altreitalie* 2 (1989): 3–35; see also Piero Gastaldo, "Gli americani di origine italiana: Chi sono, dove sono, quanti sono," in *Euroamericani* (Turin: Fondazione Giovanni Agnelli, 1987), pp. 149–99.

8 Leslie Page Moch, "Foreign Workers in Western Europe: The 'Cheaper Hands' in Historical Perspective," in Jytte Klausen and Louise A. Tilly (eds), *European Integration in Social and Historical Perspective, 1850 to the Present* (Lanham, Md.: Rowman & Littlefield, 1997), p. 193.

9 Castles and Miller, *The Age of Migration*, Table 8.1.

10 Hans-Joachim Hoffman-Nowotny, "Switzerland: A Non-Immigration Immigration Country," in Robin Cohen (ed.), *The Cambridge Survey of World Migration* (Cambridge: Cambridge University Press, 1995).

11 Christiane Harzig, unpublished paper in possession of author. Germany has recently changed its laws to facilitate naturalization of migrants' children when they remain in the country.

12 On relations between Swiss and Italian workers, see Alvaro Aretini, *Gli svizzeri e gli emigranti italiani: Il comportamento degli svizzeri e degli italiani raccontato da un emigrante* (Zurich: Ferrari, 1963).

13 Maria Rosaria Stabili, "Da sfruttati a sfruttatori; Italiani e mapuches in Capitan Pastene, Cile, 1905–1940," in Vanni Blengino, Emilia Franzina, and Adolfo Pepe (eds), *La riscoperta delle Americhe; Lavoratori e sindacato nell'emigrazione italiana in America Latina 1870–1970* (Milan: Teti Editore, 1993), pp. 291–310. For a somewhat similar argument about the much smaller group of Italians in South Africa, see Chiara Ottaviano, "Fortune, travagli e privilegi dei biellesi in Sud Africa," in Valerio Castronovo (ed.), *L'emigrazione biellese fra otto e novecento* (Milan: Electa, 1987), vol. 2, pp. 243–91. For possible connections between the Latin American experience and the growth of imperialist sentiments in Italy in the 1880s and 1890s, see also Piero Brunello, "Indios e coloni nel Brasile meridionale

(Rio Grande Do Sul e Santa Catarina): 1875–1914," in Blengino *et al.*, *La riscoperta delle Americhe*, pp. 311–25.

14 Jose C. Moya, *Cousins and Strangers; Spanish Immigrants in Buenos Aires, 1850–1930* (Berkeley: University of California Press, 1998), pp. 345–46. On pro-immigrant sentiment in Argentina, see Tulio Halperin-Donghi, "Para que la inmigracio? Ideologia y politica inmigratoria y aceleracion del proceso modernizador: el casa argentino (1814–1914)," *Jahrbuch für Geschichte von Staat, Wirtschaft und Gesellschaft Lateinamerikas* 13 (1976): 437–89; Fernando Devoto, "Acerca de la construccion de la identidad nacional en un pais de inmigrantes, el caso Argentino (1850–1930)," in *Historia y presente en America Latina, Conferencias Pronunciadas en el Centre Cultural Bancaixa* (Valencia, 1996), pp. 95–126.

15 Hilda Sabato, *La política en las calles: Entre el voto y la movilización, Buenos Aires, 1862–1880* (Buenos Aires: Sudamericana, 1998); see also Torcuato A. Di Tella, "El impacto inmigratorio sobre el sistema politico argentino," *Estudios migratorios Latinoamericanos* 4, 12 (1989): 218–22; David Rock, *Politics in Argentina, 1890–1930* (Cambridge: Cambridge University Press, 1975), pp. 16f. Argentina introduced universal manhood suffrage in 1912 — at the same time as Italy.

16 Quoted in Samuel L. Baily, *Immigrants in the Lands of Promise* (Ithaca, N.Y.: Cornell University Press, 1999), p. 78; Samuel L. Baily, "Sarmiento and Immigration: Changing Views on the Role of the Immigrant in the Development of Argentina?," in Joseph T. Criscenti (ed.), *Sarmiento and his Argentina* (Boulder, Colo.: L. Rienner, 1992), pp. 131–42; Ana Cara-Walker, "Cocoliche: The Art of Assimilation and Dissimulation among Italians and Argentines," *Latin American Research Review* 22, 3 (1987): 37–67.

17 Besides Baily, *Immigrants in the Lands of Promise*, Table 33, see Nora Pagano and Mario Oporto, "La conducta endogámica de los grupos inmigrantes: Pautas de los italianos en el barrio de La Boca en 1895," in Devoto and Rosoli (eds), *L'Italia nella società argentina*, pp. 90–101; Samuel L. Baily, "Marriage Patterns and Immigrant Assimilation in Buenos Aires, 1882–1923," *Hispanic American Historical Review* 60 (1980): 32–48; Eduardo José Miguez, "Il comportamento matrimoniale degli italiani in Argentina," in Gianfausto Rosoli (ed.), *Identità degli italiani in Argentina* (Rome: Studium, 1993), pp. 81–106; Carina Silberstein, "Inmigracion y seleccion matrimonial: El caso de los italianos en Rosario (1870–1910)," *Estudios migratorios Latinoamericanos* 6, 18 (1991): 161–90.

18 Robert Foerster, *The Italian Emigration of Our Times* (New York: Russell & Russell, 1968), p. 274.

19 Useful sources on this transition are Rock, *Politics in Argentina*, and Daniel James, *Resistance and Integration: Peronism and the Argentine Working Class, 1946–1976* (New York: Cambridge University Press, 1978).

20 A good starting place on a complex issue is Torcuato S. Di Tella, "Il ruolo dell'immigrazione di massa nella formazione del corporativismo argentino," in Rosoli, *Identità degli italiani in Argentina*, pp. 107–28. See also Mario C. Nascimbene, *Los Italianos y la integracion nacional; Historia evolutiva del la colectividad italiana en la Argentina (1835–1965)* (Buenos Aires: America Latina, 1986).

21 Angelo Trento, "Italianità in Brazil: A Disputed Object of Desire," in Lydio Tomasi, Piero Gastaldo, and Thomas Row (eds), *The Columbus People: Perspectives in Italian Emigration to the Americas and Australia* (New York and Turin: Center for Migration Studies and Fondazione Giovanni Agnelli, 1994), pp. 267–8.
22 E. Bradford Burns, *A History of Brazil*, 3rd edn (New York: Columbia University Press, 1993), Table 7.1, p. 317.
23 The fullest studies of France's immigrants are Gérard Noiriel, *Le Creuset français; Histoire de l'immigration XIXe–Xxe siècles* (Paris: Seuil, 1988). See also Yves Lequin (ed.), *La Mosaïque, France: Histoire des étrangers et de l'immigration en France* (Paris: Larousse, 1988); Gary Cross, *Immigrant Workers in Industrial France* (Philadelphia, Pa.: Temple University Press, 1983).
24 Eugen Weber, *Peasants into Frenchmen: The Modernization of Rural France, 1870–1914* (Stanford, Calif.: Stanford University Press, 1976).
25 These can be traced in Italian guides, e.g.: *Il trattamento degli stranieri nei vari paesi* (1926); *L'italiano all'estero* (1934); these are also available in the 1950s, see review in *Italiani nel mondo* 11, 5 (March 1955), p. 16.
26 Gérard Noiriel, *The French Melting Pot; Immigration, Citizenship, and National Identity* (Minneapolis: University of Minnesota Press, 1996), p. xviii.
27 James R. Barrett and David Roediger, "In-between Peoples: Race, Nationality and the 'New Immigrant' Working Class," *Journal of American Ethnic History* 16 (1997): 3–44; see also Rudolph J. Vecoli, "Are Italians White?," *Italian Americana* 12 (Summer 1995): 149–61. The following work was not yet available to me as I was writing: David Richards, *Italian American: The Racializing of an Ethnic Identity* (New York: New York University Press, 1999).
28 Salvatore J. La Gumina (ed.), *WOP! A Documentary History of Anti-Italian Discrimination in the United States*, 2nd edn (Toronto: Guernica, 1998).
29 Ronald H. Bayor, *Fiorello La Guardia: Ethnicity and Reform* (Arlington Heights, Ill.: Harlan Davidson, 1993): Thomas Kessner, *Fiorello H. La Guardia and the Making of Modern New York* (New York: McGraw-Hill, 1989).
30 A good case study of the 1930s is Stefano Lusconi, "Machine Politics and the Consolidation of the Roosevelt Majority: The Case of Italian Americans in Pittsburgh and Philadelphia," *Journal of American Ethnic History* 15, 2 (1996): 32–59. For an overall assessment of Italian-Americans in politics, see Michael Parenti, *Ethnic and Political Attitudes: A Depth Study of Italian Americans* (New York: Arno Press, 1975). On the persisting significance of ethnicity in New York politics, see Nathan Glazer and Daniel Patrick Moynihan, *Beyond the Melting Pot: The Negroes, Puerto Ricans, Jews, Italians and Irish of New York City* (Cambridge, Mass.: Massachusetts Institute of Technology Press, 1968).
31 Baily, *Immigrants in the Lands of Promise*, pp. 151–2.
32 Richard Alba, *Italian Americans: Into the Twilight of Ethnicity* (Englewood Cliffs, N.J.: Prentice-Hall, 1985).
33 On suburbanization, see Salvatore LaGumina, *From Steerage to Suburb: Long Island Italians* (New York: Center for Migration Studies, 1990).
34 A general study is Michael Novak, *The Rise of the Unmeltable Ethnics: Politics and Culture in the Seventies* (New York: Macmillan, 1973).

35 Gianfausto Rosoli, "Italian Migrants and the Transformation of a Myth," in Dirk Hoerder and Horst Rössler (eds), *Distant Magnets: Expectations and Realities in the Immigrant Experience, 1840–1930* (New York and London: Holmes & Meier, 1993), pp. 232–3.
36 Jerre Mangione and Ben Morreale, *La Storia; Five Centuries of the Italian American Experience* (New York: HarperCollins, 1992), ch. 27.
37 Dino Cinel, *The National Integration of Italian Return Migration 1870–1929* (Cambridge: Cambridge University Press, 1991), p. 233; other general interpretations of Italian-American life include Luciano J. Iorizzo and Salvatore Mondello, *The Italian-Americans* (New York: Twayne, 1971); Joseph Lopreato, *Italian Americans* (Austin: The University of Texas Press, 1970); Humbert S. Nelli, *From Immigrants to Ethnics: The Italian Americans* (New York: Oxford University Press, 1983). A study that particularly emphasizes the isolation of Italian Americans from the mainstream, and their inward orientation, is Richard Gambino, *Blood of my Blood; The Dilemma of Italian Americans*, 2nd edn (New York: Guernica, 1996).
38 Robert F. Harney, "Italophobia: An English-speaking Malady?," *Polyphony* 7 (1985).
39 Bruno Ramirez, "Ethnic Studies and Working-class History," *Labour/Le Travail* 19 (Spring 1987): 45–8; see also Bruno Ramirez, *The Italians in Canada* (Ottawa: Canadian Historical Society, 1990). For a recent critique, see Neil Bissondath, *Selling Illusions: The Cult of Multiculturalism in Canada* (Toronto: Penguin Books Canada, 1994).
40 Stephen Castles (ed.), *Australia's Italians: Culture and Community in a Changing Society* (Sydney: Allen and Unwin, 1992); see also Stephen Castles, "Italians in Australia: the Impact of a Recent Migration on the Culture and Society of a Postcolonial Nation," and Ellie Vasta, "Cultural and Social Change: Italian-Australian Women and the Second Generation," in Tomasi *et al.*, *The Columbus People*, pp. 342–67 and 406–25.
41 On this point, the key works are Herbert Gans, "Symbolic Ethnicity: The Future of Ethnic Groups and Cultures in America," *Ethnic and Racial Studies* 2 (1979): 1–10; Mary Waters, *Ethnic Options: Choosing Identities in America* (Berkeley: University of California Press, 1990); Richard Alba, *Ethnic Identity: The Transformation of White America* (New Haven, Conn.: Yale University Press, 1990); a sensible response is George E. Pozzetta, "What Then is the European American, This New Man?: Ethnicity in Contemporary America," *Altreitalie* 6 (1991): 114–17.
42 Nathan Glazer, *We Are All Multi-Culturalists Now* (Cambridge, Mass.: Harvard University Press, 1997). For a useful comparison (if somewhat different from the one offered here), Sneja Marina Bunew, "Multicultural Multiplicities: U.S. Canada, Australia," in *Cultural Studies: Pluralism and Theory* (Melbourne: University of Melbourne Press, 1993), pp. 51–65.
43 Useful background in English is in Stephen Gundle and Simon Parker (eds), *The New Italian Republic; From the Fall of the Berlin Wall to Berlusconi* (London: Routledge, 1996); Patrick McCarthy, *The Crisis of the Italian State; From the Origins of the Cold War to the Fall of Berlusconi* (New York: St. Martin's Press, 1995).
44 The source of this distinction is Robert Putnam, *Making Democracy Work: Civic Traditions in Modern Italy* (Princeton, N.J.: Princeton University Press, 1993).

45 For current discussions of the Italian nation, see Paolo Casini, "Cultura nazionale e dimensione internazionale," in *Cultura nazionale, culture regionali, comunità italiane all'estero* (Rome: Istituto della Enciclopedia Italiana, 1988), pp. 21–8; G.E. Rusconi, *Se cessassimo di essere una nazione* (Bologna: Il Mulino, 1993); Giorgio Calcagno (ed.), *Bianco, rosso e verde: L'identità degli italiani* (Rome: Laterza, 1993); Ernesto Gali della Loggia, *L'identità italiana* (Bologna: Il Mulino, 1998).
46 "Many Mountains to Climb," *The Economist*, November 8, 1997, p. 4.
47 For Argentina, assimilation theory begins with the work of Gino Germani. In English, see his "Mass Immigration and Modernization in Argentina," in Jorge Dominguez (ed.), *Race and Ethnicity in Latin America* (New York: Garland Publishing, 1994); Arnd Schneider, "The Two Faces of Modernity: Concepts of the Melting Pot in Argentina," *Critique of Anthropology* 16 (1996): 173–98.
48 *New York Times*, "Week in Review," May 24, 1998.
49 Mario Anselmi, "Presenza italiana nel mondo," *Italiani nel mondo* 21, 11 (June 1965).
50 Donald Horowitz, "Immigration and Group Relations in France and America," in Donald Horowitz and Gérard Noiriel (eds), *Immigrants in Two Democracies: French and American Experience* (New York: New York University Press, 1992), p. 7.
51 Alec G. Hargreaves, *Immigration, "Race" and Ethnicity in Contemporary France* (London: Routledge, 1995).
52 On France, see Nancy Green, "'Le Melting Pot': American Concept, French Reality," *Journal of American History* (forthcoming); see also Gérard Noiriel, "Immigration and National Memory in the Current French Historiography," *IMIS-Beiträge* 10 (1999): 39–55. On Argentina, see Fernando J. Devoto, "Discutiendo las relaciones entre los italianos y la sociedad argentina; Problemas de modelos y fuentes," in *Estudios sobre la emigración italiana a la Argentina en la segunda mitad del siglo XIX* (Sassari: Ed. Scientifiche Italiane, 1991) and Luigi V. Favero, "Mechanisms of Adaptation and Integration of Italian Immigrants in Argentina: From Social Spaces to the Interpretative Paradigms of Ethnic Identity," in Tomasi *et al.*, *The Columbus People*, pp. 113–24; Arnd Schneider, "Discours sur l'altérité dans l'Argentine moderne," *Cahiers Internationaux de Sociologie* 55 (1998): 341–60.
53 Axel Schulte, "Multikulturelle Gesellschaft: Chance, Ideologie oder Bedrohung?," *Aus Politik und Zeitgeschichte* 23–4 (1990): 3–15.
54 Nancy Green, "The Comparative Method and Poststructural Structuralism — New Perspectives for Migration Studies," *Journal of American Ethnic Studies* 13 (Summer 1994): 3–22.
55 *Italiani* (July/August 1994).
56 Romolo Gandolfo, "Inmigrantes y politica en Argentina: La Revolucion de 1890 y la campana en favor del la naturalizacion automatica de residentes extranjeros," *Estudios migratorios Latinoamericanos* 6, 17 (1991): 23–54. See also Giuseppe Lupis, "Gli italiani in Argentina e il problema della cittadinanza," *Italiani nel mondo* 21, 3 (February 1955).
57 Moya, *Cousins and Strangers*, p. 489.
58 For the U.S. see Baily, *Immigrants in the Lands of Promise*, pp. 74, 83–5; on France, see Gabaccia and Ottanelli, "Diaspora or International Proletariat?"

59 I. Baldelli (ed.), *La lingua italiana nel mondo* (Rome: Istituto della Enciclopedia Italiana, 1987).

60 For the U.S., where the association of Italians with crime is strongest (probably because a higher proportion than elsewhere originated in Italy's South), see Humbert S. Nelli, *The Business of Crime: Italians and Syndicate Crime in the United States* (Chicago: University of Chicago Press, 1976); Francis A.J. Ianni, *A Family Business: Kinship and Social Control in Organized Crime* (New York: New American Library, 1972); Thomas M. Pitkin, *The Black Hand: A Chapter in Ethnic Crime* (Totowa: Rowman and Littlefield, 1977).

61 Mangione and Morreale, *La storia*, p. 241.

62 From a growing literature, see Carlos E. Cortes, "The Hollywood Curriculum on Italian Americans; Evolution of an Icon of Ethnicity," in Tomasi *et al.*, *The Columbus People*, pp. 89–108.

63 These are mainly my personal observations. See the very lively exchange on food as a controversial source of Italian identity on the internet discussion group H-ItalAm, in 1996–7, and the special issue on Italian-American food in *Italian Americana* (1998).

64 The best introduction to Italian-American culture is Pellegrino D'Acierno, *The Italian American Heritage; A Companion to Literature and Arts* (New York: Garland Publishing, 1999). On the connection between food, patriarchy, and women, see Daniel S. Golden, "Pasta or Paradigm: The Place of Italian American Women in Popular Film," *Explorations in Ethnic Studies* 2 (1979): 3–10. A review essay on women, family, and patriarchy among Italian-Americans, is Donna Gabaccia, "Italian Immigrant Women in Comparative Perspective," in Tomasi *et al.*, *The Columbus People*, pp. 391–405.

65 From a huge literature, the best starting place remains Virginia Yans McLaughlin, "Patterns of Work and Family Organization: Buffalo's Italians," *Journal of Interdisciplinary History* 2 (1971): 299–314.

66 In fact, in this regard, they differ little from Jewish Americans or Americans of south Slav descent; see Corinne Krause, "Urbanization without Breakdown: Italian, Jewish and Slavic Women in Pittsburgh, 1900 to 1945," *Journal of Urban History* 4 (1978): 291–305.

67 For a recent discussion in Italy, see Silvio Lanaro, *Patria: Circumnavigazione di un'idea controversa* (Venice: Marsilio, 1996). While widely noted of Italian-Americans, localism (like the concept of campanilismo) has not received systematic scholarly exploration.

68 *Italiani nel mondo* 21, 23 (December, 1975).

69 In the U.S., by contrast, it was simply lost (or destroyed). See Philip V. Cannistraro (ed.), *The Lost World of Italian-American Radicalism* (New York: SUNY Press, forthcoming).

INDEX

2600